BEDROCK

BEDROCK

Writers on the Wonders of Geology

Edited by Lauret E. Savoy, Eldridge M. Moores, and Judith E. Moores

Foreword by Gordon P. Eaton

8/16/11

Our Earth —
the only home for
us in the whole wide
universe.
Thank you to you
help teaching
us how to live
closer to our
planet — by
getting our
hands dirty +
growing our
own food!

Thank you.

Judy

TRINITY UNIVERSITY PRESS
San Antonio

The publisher gratefully acknowledges the generous support of the Geological Society of America Foundation toward the publication of this book.

Published by Trinity University Press
San Antonio, Texas 78212

Cover design by Erin K. New
Book design by BookMatters, Berkeley

⊛ The paper used in this publication meets the minimum requirements of the American National Standard for Information Sciences—Permanence of Paper for Printed Library Materials, ANSI Z39.48-1992.

Library of Congress Cataloging-in-Publication Data

Bedrock : writers on the wonders of geology / edited by Lauret
    E. Savoy, Eldridge M. Moores, and Judith E. Moores.
        p.   cm.
    ISBN-13: 978-1-59534-022-1 (hardcover : alk. paper)
    ISBN-10: 1-59534-022-X (hardcover : alk. paper)
    ISBN-13: 978-1-59534-023-8 (pbk. : alk. paper)
    ISBN-10: 1-59534-023-8 (pbk. : alk. paper)
        1. Nature—Literary collections.   2. Geology—Literary
    collections.   I. Savoy, Lauret E.   II. Moores, Eldridge M.,
    1938–   III. Moores, Judith E., 1943–
PN6071.N3B43   2006
808.8'036—dc22   2005030817

10 09 08 07 06   5 4 3 2

Contents

Acknowledgments

This anthology is a product of wonderful collaboration. It began as a suggestion from Robert Fuchs of the Geological Society of America Foundation for a sequel to *The Art of Geology* (1988, GSA Special Paper 225). As such it originally was conceived as a collection of photographs and writings, during the preparation and production of which, the project evolved into an anthology of writings. We wish to thank Robert Fuchs for suggesting the project.

We owe our deepest gratitude to Gordon Eaton, former Director of the U. S. Geological Survey, for his belief in the project and generous financial contribution. We also give very special thanks to John Elder, Jack Hicks, Louis Owens, John Lemly, and Jon Olsen (Director of Publications of the Geological Society of America), without whose valuable insights and encouragement this book would have remained an idea.

We are grateful to colleagues and friends who offered their guidance as we searched for and selected diverse writings and, originally, photographs. The following individuals suggested authors or selections: David Applegate, Vic Baker, Marcia Bjornerud, Kathy Cashman, Jane Crosthwaite, Ted Daeschler, Alison Deming, Paul Doss, Robert Dott, John Elder, Deborah Gorlin, Naomi Horii, Linda Ivany, Irene Klaver, John Lemly, David Leveson, Michelle Markley, Carl Mendelson, Lynn Miller, Marli B. Miller, Carl Mitcham and Robert Frodeman, Naomi Oreskes, Alison (Pete) Palmer, Orrin Pilkey, Paul Pinet, Paul H. Reitan, Karen Remmler, Frank Rhodes and Bruce Malamud, David Robertson, Jill Schneiderman,

Michelle Stephens, Roger D. K. Thomas, Robert Torrance, Sarah Trainor, Stephen Trimble, Christine Turner, and E-An Zen. This book is without doubt a more interesting one because of their generous help. Ann Zwinger, John Calderazzo, Jane Hirshfield, John McPhee, and Gary Snyder showed great enthusiasm for the project and for having their work included. Don Easterbrook and Richard Woldendorp have also provided the use of their photographs.

Our publisher, Barbara Ras, was an inspired collaborator and guide throughout the process. Thank you, Barbara, and thank you, BookMatters! We are also grateful to John Elder for his review of the first draft of the book, and Blake Edgar and Malcolm Margolis for early discussions on publishing.

For assistance in typing and scanning the materials, we thank Amie DeGrenier and Mount Holyoke College students Lisa Hupp, Clarisse Hart, Leigh Dumville, Gaytri Bhatia, Leise Jones, Jessica Whiteside, Ramona Smith, and Jennifer Peacock.

Finally, our thanks go as well to Steven Roof, Elizabeth Hemley, Alison Deming, and other friends for their encouragement.

Foreword

You hold in your hands an impressive volume, a unique anthology on the literature and art of geology. It was assembled by a dedicated and talented group of editors: Lauret Savoy, professor of geology and environmental studies at Mount Holyoke College; Eldridge Moores, internationally respected senior earth scientist at the University of California, Davis; and Judith Moores, wife of Eldridge, a gifted woman in her own right, trained in biology, literature, and art.

The book speaks to the profound influence of Earth's landscapes and physical processes on human perceptions and cultures. It does so through the use of splendid literature from original sources extending from ancient times to the present and from observers, writers, and thinkers from around the globe.

Anne Spirn noted in her 1998 book, *The Language of Landscape*, that "Landscapes were the first human texts." They remain so still, basic texts to which we have simply added many other sources of knowledge and means of instruction.

The importance of this influence of landscape on human cultures and thought has been made clear in another place, a sign hanging near the entrance to the magnificent First Peoples Gallery of the Royal British Columbia Museum in Victoria, B.C. The sign instructs us that "In order to survive, humans must provide for their material, emotional and intellectual needs. These are satisfied by culture, a complex system that includes tools, language, arts and beliefs. Cultures vary because they must be compatible with their supporting environments. Thus different climates, terrains, and sources of food evoke different cultural responses."

Landscapes and the geology that underlies them are profound in their influence on us all. For some, this influence is aesthetic; for others, spiritual. For yet others, landscapes fire the imagination, inspire, or are a source of endless curiosity and the object of investigation. However landscapes may touch you, dear reader, they exert their influence on you in ways that are, at the same time, both subtle and powerful.

Gordon P. Eaton
12th Director (retired)
U.S. Geological Survey

A rain of black rocks   out in space
onto deep blue ice   in Antarctica
nine thousand feet high   scattered for miles.

Crunched inside   yet older matter
from times before our very sun

—Gary Snyder, "Yet Older Matters"
(from a conversation with Eldridge Moores
and Kim Stanley Robinson)

Introduction

How do we "know" Earth? How do we remember and record our living in this world? From the beginning, human beings have been influenced by the rocks, soil, rivers, mountains, deserts, and forests where we evolved. As Anne Spirn notes in *The Language of Landscape*, "Humans touched, saw, heard, smelled, tasted, lived in, and shaped landscapes before the species had words to describe what it did. Landscapes were the first human texts, read before the invention of other signs and symbols."[1] The connection has always been reciprocal. From ancient times on, Earth's physical faces have imprinted themselves in cultural traditions; and so, too, humankind has changed these faces through agriculture, settlement, resource use, and industry.

For over four centuries geology, the scientific discipline that seeks to understand Earth, has provided knowledge of earth materials, properties, processes, and history based increasingly on analytic description, abstraction, and specialization. So particularized had this knowledge become that in *A Sense of the Earth* (1972), David Leveson was led to ask whether geology and geologists could even recognize—let alone interpret—the crucial relationships between societies and Earth, what he called the larger "geologic experience." At the time of Leveson's writing, the "Plate Tectonic Revolution" was well under way, uniting geology's subfields in an all-embracing theoretical framework. Although most geoscientists study Earth as complex chemical, physical, and biological systems interacting through the planet's long history, that broader focus still invites exploration by us all.

Inquiry into a larger human experience beyond specialized science enlarges our knowledge of Earth. The histories and direction of exploration, settlement, and habitation tell of continuous encounters between landscape and culture. Oral tradition, art, literature, maps, scientific theories—stories of imagination and knowledge—are all attempts to describe and understand Earth and humankind's place on it. Yet how many of us reflect on how we tell stories of the physical world and about ourselves in the world, and then attempt to live by such stories? These representations wield great power, determining in large measure, in our thoughts and actions, how we use the land.

The word *geology* comes from two Greek roots: *geo-* (Earth) and *-logia* (discourse, study of). This collection takes that definition broadly and invites inquiry beyond the discipline's customary boundaries to explore encounters with Earth's landscapes and geological processes in human imagination, perception, and experience. Rather than catalog geology's historical development or current issues, we offer a thematic, interdisciplinary sampling of writings—most previously published—beyond and within the geosciences. Selections of nonfiction, fiction, and poetry from many historical periods look at Earth from myriad perspectives: cultural, literary, artistic, scientific, social.

We want readers to reflect on the many aspects of the "geologic experience": the complex connections between Earth and humans in time and space; the impact of earth processes and features on the lives of individuals and societies; the range of narratives that shape our sense of Earth, past and present, and thus offer a context for the field of geology. We want readers to enter *Bedrock* with a large sense of the human possibilities of that word. "Bedrock" refers to a secure foundation, to basic principles, to an elemental essence—in addition to the solid rock beneath us and soil.

There is great diversity in how humankind has known and responded to Earth. *Bedrock* includes works from around the world, written from many positions—traditional and indigenous as well as Western scientific. We seek to avoid either generalizing or privileging some types of knowledge over others. This collection contains the words of artists and anthropologists, traditional elders and philosophers, novelists and poets, aviators, naturalists, and scientists of many flavors. Selections describe earthquakes felt in Japan (in 1185) and San Francisco (in 1906), close views of erupting volcanoes, reflections on patterns in stone and sculpture, deserts in Africa and America, the flow of rivers and ice, nuclear landscapes in the American Southwest, soil erosion in the 1930s American Dustbowl, and accounts of Earth in cultural memory and tradition. These varied samples reveal the interconnectedness of process and response, material and structure, fragment and history.

Specific landscape features or materials—rivers, mountains, or rock—and geologic processes—volcanic eruptions, earthquakes, or erosion—recur in writings about Earth. The book is organized in a way that provides a coherent sequence, from earth material and time, to examples of these landscape processes and features, to human reflections on Earth. But the sections also stand alone and can be read in any order. The first two parts present framing views of rock and stone as elemental materials, and of Earth's antiquity and "deep time." The following parts on earthquakes, volcanic eruptions, and mountain uplift each consider land-shaping effects of Earth's dynamic internal processes driven by plate tectonics. The next parts treat the Sun- or climate-driven work of water, ice, and wind—what geologists call surface processes and landscapes. The final section reflects on the human place on Earth.

Although far from comprehensive, this collection raises a fundamental question: How can and should we relate to Earth, locally, regionally, and globally? How we—all human beings—live in the world requires an informed understanding of Earth's environments and ecological processes, and of how humankind across traditions, past and present, has experienced this world. Human actions have always occurred in their cultural, geological, and ecological contexts: if we—all of us—understand that past and present more fully, we might better guide ourselves to an environmentally sound future. We hope this book, by bridging specialized science and ordinary existence, will help us better sense and live the larger "geologic experience" encompassing all our lives.

NOTE

1. Anne Spirn, *The Language of Landscape* (New Haven: Yale University Press, 1998).

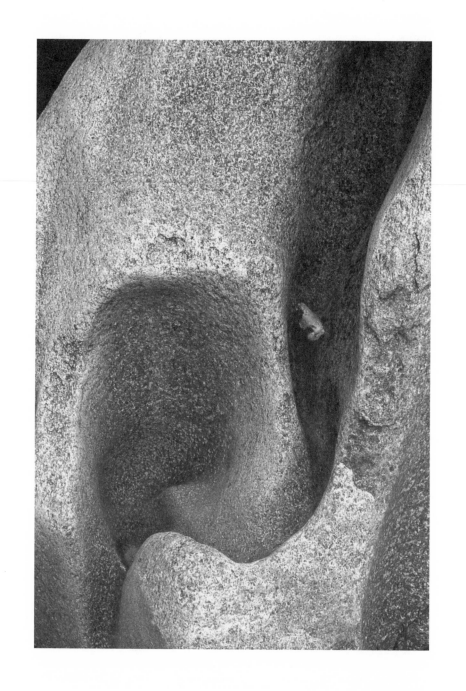

1 Of Rock and Stone

Go inside a stone.
That would by my way.
Let somebody else become a dove
Or gnash with a tiger's tooth.
I am happy to be a stone.

From the outside the stone is a riddle:
No one knows how to answer it.
Yet within, it must be cool and quiet
Even though a cow steps on it full weight,
Even though a child throws it in a river;
The stone sinks, slow, unperturbed
To the river bottom
Where the fishes come to knock on it
And listen.

I have seen sparks fly out
When two stones are rubbed,
So perhaps it is not dark inside after all;
Perhaps there is a moon shining
From somewhere, as though behind a hill—
Just enough light to make out
The strange writings, the star-charts
On the inner walls.

—Charles Simic, "Stone"

LET'S BEGIN WITH FOUR QUESTIONS: DO ROCK AND STONE SPEAK OF history and origins? What stories do they tell? How do we, as human beings, acquire meaning through observing and living on the material Earth? Can we read ourselves-in-the-universe through stone?

The paths of human cultures, and our built environments, have taken direction and shape from the physical basis rock provides in the lay of the land and in the yielding presence of extractable "resources." Each pebble or stone, each outcrop, each grain of sand, also is a relic of former worlds—of volcanic eruptions, of mountains uplifted and eroded, of Earth's interior intruded and metamorphosed. A geologist's prize is to read Earth history, process, and structure from what rock retains, after weathering and erosion.

If we look closely, we find that most rocks are assemblages of naturally occurring minerals with specific compositional and physical characteristics. Most minerals are mainly combinations of the elements oxygen, silicon, and aluminum, the three most abundant elements in Earth's crust. Other elements find homes in these structures that fit their atomic size. Although more than two thousand different minerals have been described to date, only about a dozen minerals (such as quartz, feldspar, olivine, and pyroxene) make up most of the rocks of Earth's crust.

But rock and stone invite multiple interpretations beyond the sciences— vocabularies of different cultural experiences, and literary and artistic expressions, have taken shape and meaning from this mineral world. For example, Louise Erdrich notes that in the language of her people, the Anishinabe or Ojibwe, the "word for stone, *asin,* is animate. After all, the preexistence of the world according to Ojibwe religion consisted of a conversation between stones. People speak to and thank stones in the sweat lodge, where the asiniig are superheated and used for healing. They are addressed as grandmothers and grandfathers. Once I began to think of stones as animate, I started to wonder whether I was picking up a stone or it was putting itself in my hand. Stones are no longer the same as they were to me in English."[1]

The writings in this section present many different stories of rock and stone. Included here are works in poetry, fiction, and nonfiction that reflect on elemental connections of place and culture for traditional societies; on pattern, texture, and design in stone and rock; on rock as indicator and framework of our experience on Earth. Also included are stories of the search for a geologic understanding of Earth process, structure, and history, and the ultimate inability of human analysis and language to grasp and articulate the unknowable.

This section begins with an excerpt from Hikaru Okuizumi's novel *The Stones Cry Out* (1993), winner of the Akutagawa Prize, Japan's prestigious literary award. Okuizumi notes that "even the smallest stone in a riverbed has the entire history of the universe inscribed upon it." The excerpt considers how an encounter with a dying lance corporal—a geologist—on the Philippine island of Leyte in the Second World War led a veteran to become "a fanatic of stones."

From *A Sense of the Earth* (1972), geologist David Leveson reflects on the "innocence of rock." Then Luci Tapahonso tells stories of the rocky lands in Diné (Navajo) country in her essay "Ode to the Land: The Diné Perspective."

British archaeologist and writer Jacquetta Hawkes (1910–96) describes the "continued presence of the past" in her award-winning book *A Land* (1951). The selection presents a "digression" on the human connection to rock and soil, with examples in sculpture and architecture. Hawkes notes that "Life has grown from the rock and still rests upon it; because men have left it far behind, they are able consciously to turn back to it."

The land is a key participant in *A Passage to India* (1924), the masterwork and last novel of English author and critic E. M. Forster (1879–1970). From his account of India during its movement for independence from British rule, we include a passage on caves, rock, and time.

In "The Relic Men" from *The Night Country,* Loren Eiseley (1907–77), Nebraska native, anthropologist, and former curator of Early Man at the University of Pennsylvania Museum, relates his encounter with a petrified woman and "the pathos of a man clinging to order in a world where the wind changed the landscape before morning."

An essay by contemporary natural history writer-artist Ann Zwinger, "A Question of Conglomerate," follows. Zwinger has asked elsewhere, "Why cannot we, as teachers and writers, give a cleaner, clearer sense of this earth that cradles us all, affects us all, directs us all, a sense that geology is the logical essence of all our histories and ourselves?"

The section ends with the poem "Rock" by Jane Hirshfield (1953–). Of Hirshfield's poetry, W. S. Merwin has noted, "These are poems of space, air, and a remarkable precision of observation and revealed feeling."

NOTE

1. Louise Erdrich, *Books and Islands in Ojibwe Country* (Washington D.C.: National Geographic Directions, National Geographic Society, 2003).

Hikaru Okuizumi

FROM *The Stones Cry Out* (1993)
  translated from the Japanese by James Westerhoven

Even the smallest stone in a riverbed has the entire history of the universe inscribed upon it. The reason Tsuyoshi Manase became a fanatic of stones can be traced back to words spoken to him by a dying man during the Second World War, in mid-December 1944, in a cave in the middle of the tropical forest above the Bay of Carigara in northern Leyte.

The man was wasted by malnutrition and amoebic dysentery, and his face resembled a skeleton of wires covered with parchment. Only his eyes moved, restlessly. These eyes he now fixed on Manase. With his emaciated fleshless fingers that seemed more like roots to Manase, the man picked up a stone from the ground.

This should be classified as green chert, he said in a magisterial tone, as if he were addressing a group of students. The cave was formed when bedrock from the Paleozoic era rose to the surface and was eroded by the sea. Later, during the Quaternary era, the sea withdrew and left the cave in the midst of jungle. Thus the walls around them were probably full of fossilized marine organisms. If you were to examine this little piece of rock under a microscope, the man informed Manase, you would be sure to find radiolarians and the like. His lecture continued more or less as follows:

"You normally don't pay much attention to the stones you see by the side of the road, do you? Oh, perhaps if they're stones you can use for your garden, or your house, say, but in general you don't give much thought to them. You just think of them as meaningless objects scattered in the mountains, rivers, and fields. Even if they're in the way, it doesn't occur to you that they might be worth picking up and studying. Well, you're wrong, you know. Even the most ordinary pebble has the history of this heavenly body we call earth written on it. For instance, do you know how rocks are formed? Rocks are formed when red-hot magma cools and solidifies; rock erodes under the influence of wind and weather on the surface of the earth. That's how you get stones. Stones are eventually ground into sand, sand into soil; then stones and sand and soil are carried away by streams and settle on the bottom of lakes, fens, or the sea, where they once again harden into rock. That rock crumbles and changes back into stones and sand and soil, or it may be pushed deep beneath the surface of the earth and, under the influence of heat and tremendous pressure, reborn as rock, in all shapes and sizes; or sometimes it melts into magma and returns to its origins. The form of minerals is never static, not for a second; on the contrary, it undergoes constant change. All matter is part of an unending cycle. You know of course that even the continents actually move, though at an imperceptibly slow pace.

"What I'm trying to say is, the tiny pebble that you might happen to pick up during a walk is a cross-section of a drama that began some five billion years ago, in a place that would later come to be called the solar system—a cloud of gas drifting idly through space, growing denser and denser until after countless eons it finally gave birth to this planet. That little pebble is a condensed history of the universe that keeps the eternal cycle of matter locked in its ephemeral form."

Manase had spent a year and a half as a prisoner of war in the camp at Calanban. Only after he was released and had settled in the village where his mother and father had fled from the war had he begun to dwell on the words the man had spoken as he lay stretched out on his bed of rocks. What immediately struck Manase was not the words he had heard in that cave with its evil stench, but the maggots. As a person's flesh and fat waste away, the eyes begin protruding from the skull, which explains why a starving person's eyes seem unnaturally large. If a skeleton had eyes, they would be nothing if not conspicuous. Since the eyes of people weak with sickness generally lie motionless in their sockets, Manase thought it strange that this man's pupils moved relentlessly as he talked. When Manase looked more closely, he saw maggots. The man's eyeballs were swarming with them. Not that maggots were a rare sight in the cave—all around lay dozens of rotting corpses with maggots crawling beneath their skin, and Manase himself had a sore on his knee filled with them—but this was the first time he had seen maggots squirm in the eyes of a breathing, talking human being.

But what impressed Manase most was the intimate, fatherly tone in which the man addressed him. No one had spoken to Manase in such a friendly way from the moment he had been drafted in early January 1943, when he joined the regiment in Yamanashi, until he reached the front, near the Pacific. This man was not only a lance corporal and therefore Manase's superior, but more important a veteran several years his senior. That such a man would suddenly address him so warmly threw Manase into a panic. At first he did not realize the man was talking to him; when he did, he was so dumbfounded he could not keep his eyes fixed on the middle distance as military discipline required, but instead let them wander. Manase lay down beside the man and closed his eyes. The delirious lance corporal continued talking into the darkness of the cave long after Manase had turned on his side, facing away from him.

On the front, sleep often comes suddenly, in the form of deep coma, yet the words stayed with Manase. That he heard them still was less peculiar than stories he had heard in the camp, such as the vision of a lady in ancient court dress beckoning unceasingly from between the dark waves, or the jellyfish creature the size of a whale that a fighter pilot had seen floating above the clouds. Compared to such apparitions, a talking soldier hardly seemed reason to ponder the strangeness of memory. Perhaps the lance corporal's words had made their deep impression because they contrasted so poignantly with the incessant drone of airplanes and the roar of cannon on the front. Soon, Manase's recollections of the war began to fade, and all that remained fresh in his memory was the lance corporal's sermon about the stone.

Later, when Manase began to play around with stones, those words assumed a distinct shape and lodged themselves in his heart. Whenever someone asked how he began collecting stones, Manase would let his memory drift back to the time spent inside the cave on Leyte. He was never able to find the right words, and because he was not eloquent, he would reply only with a modest smile from his small, deep-set eyes and hope that would answer the question.

[. . .]

As his knowledge increased and he became more absorbed in geology, he was forced to think back to the Leyte cave and the lance corporal who had explained what was special about stones. The captain with the sword and, without any doubt, the lance corporal were the only people who had left an impression on Manase during the war. But whereas the captain's figure still seemed scorched onto his eyelids, Manase had only hazy memories of the lance corporal's face. He had already been in the cave when the group led by the captain arrived, so he probably was one of the survivors of the Kwantung army who had been transferred to the southern front in the fall of 1944. At first he had been fairly fit despite tropical dysentery, for he and Manase had foraged together. They were often separated from the others, so they spent much time in each other's company, but perhaps because of their continual fear of encountering American soldiers, Manase could not remember having a real conversation or even an opportunity to ask the other's name or where he was born. The lance corporal opened his mouth only when absolutely necessary. The sick soldiers in the cave seemed to trust him completely, and because the lance corporal always seemed to know what he was doing, Manase considered himself lucky to have teamed up with him.

What sort of person was he? He appeared fairly old, about the same age as the captain, but that in itself was nothing special, for at the Leyte front there were many veterans older than thirty. Perhaps he seemed older because of his calm and relaxed manner. He was not a student. The army was filled with people from many walks of life, so he might have been a scholar or the employee of a mining company. At any rate he was someone who had done work requiring knowledge of geology—beyond that Manase did not speculate. Later he was sorry he had not at least delivered a lock of his hair to his family, but by then the war had been over for some time.

The lance corporal's health deteriorated rapidly. Manase noticed how still he lay in his corner of the cave, and later he heard him talking about stones in the dark, knowing that these would probably be his last words. Manase had seen it often: someone still strong enough to burst into laughter one day would be still and cold the next. So Manase's heart had gone numb and no longer reacted to human life or death. Too exhausted to take any interest in events not concerning him directly, Manase was unmoved by this man's impending death.

"Stones are not only formed by magma. There are also meteorites that come flying from outer space. But the major cause is organic activity. Along with water or ice, living organisms also play an important role in the erosion process, and then their bodies in turn change into stone. Surely you know that coal is nothing but fossilized wood from ancient trees. Limestone and chert consist of the compressed skeletons of tiny organisms that collected eons ago on the bottom of the sea. Even the calcium in our own bones will eventually change into stone and be made part of the mineral cycle. That is why the tiny pebble that you pick up from a riverbed, no matter how silent and alien, is in fact your very distant relative. That pebble in your hand tells the history of the world, and you too are a part of that history, and what you discover is the way you yourself will look in the future."

The lance corporal talked on. Surely his sudden outburst was an omen that the flame of his life was about to burn out. Manase may have purposely decided to sleep next to him, for though they had seldom talked during their acquaintance, he felt respect for the man and wanted to nurse him during his last moments. [. . .]

Manase had dedicated himself to his hobby for more than ten years, reaching the point at which his geological expertise—despite the autodidact's lopsidedness—matched that of any specialist in the field. Yet the bookseller in him realized that someone who has not undergone even the most elementary academic training can hardly present himself as a scholar. He therefore separated work and pleasure, never forgetting his real profession. Business was good—possibly because the bookstore was located near the center of the city; he expanded and renovated his shop and hired a few employees, though Manase refused to leave his work to others. Every morning he bicycled faithfully to his store where he spent the day helping customers, taking orders, arranging books. But the evenings were reserved for his hobby, from the moment he returned home till the moment he went to bed.

After dinner he would pick up the mug of tea his wife had poured for him and take it to his attic. There he smoked a cigarette and sipped his hot tea, letting his eyes wander over the racks of specimens. Then he resumed his work where he had ended the night before. He switched on a lamp on his desk and began polishing a section on a glass plate. Every now and then he would hold it up against the light to check for irregularities. He polished, bit by minuscule bit, with all the care his fingertips allowed. At the slightest suspicion of roughness indicating the presence of a coarser particle in the stone, he would rinse the glass plate in a tub of water, carefully removing all the dirt with an old toothbrush, and continue polishing with another abrasive. It was exhausting work, but because he was consumed by his task, two or three hours passed in what seemed a second. One section might take from three days to a week, depending on the material, so if after all that effort the specimen came out nicely, his emotion was all the stronger for it. Once the stone's surface had acquired a satisfactory gloss, he glued the section under a cover glass, scraped off the excess Canada balsam with a pocketknife, stuck on a label with the name and site of origin, and the job was done. After placing the finished section under the microscope, he would fill a wineglass with the grape juice his father-in-law sent him each year (since he did not drink alcohol) and raise a solitary toast. Then, calmly, he switched on the microscope lamp, trying to control the beating of his heart as he brought his eye to the lens. Of course he also owned a commercial set of sections, neatly arranged in a box, but even if the stone he was about to observe might have slight flaws, he had dug it up himself and processed it entirely with his own hands, and to the eye that peered through the lens, that knowledge made all the difference.

Each rock-forming mineral has its own chemical composition and its own characteristic refractive index. Under the white light of a polarizing microscope, they display interference colors resembling those of the rainbow, in combinations so bold and variegated that they stunned him each time with the miracle of their beauty. A scarlet shaft flashing through a field of cool blue; red, green, and yellow cheerfully intermingling as on a ship in full bunting; pale turquoise enveloped in a solid ring of black-speckled gray. The fabric woven by crystals and groundmass is a breathtakingly delicate natural design. The icicles in plagioclase hang so close together that they block your field of vision; the light and dark tiles of calcite form an intricate mosaic; the snowfield of quartz extends as far as the eye can reach; rectangular spaceships of mica and feldspar fly through a black night sky past an enormous hornblende moon.

At such moments Manase always recalled the lance corporal's words in the Leyte cave. *A stone is the condensed history of the earth.* The phrase would tremble in his mind, and because he increasingly shared the same opinion, he would nod and cast another glance through the lens. The crystals lay still and motionless, hemmed in between two plates of glass, but eventually, after an endless period of time by human standards, these crystals might begin to grow again. Under heat and pressure their nondescript, expressionless, noncrystalline groundmass might even produce new crystals or perhaps break up and melt. For this solitary second the eternal cycle of matter, with its everlasting changes, had been frozen.

If he stared long enough at the mineral tapestries flickering under the microscope, the crystals seemed to possess an inner urge to grow. This urge had been forcibly repressed by some sort of magic keeping them locked inside this narrow space, but if that spell was somehow broken, would not the minerals burst into movement? Would not the crystals begin to move before his eyes—clash, intermingle, or perhaps collapse in demonstration of the infinite process of transformation? All at once the world under his eyes appeared to come to life, each mineral seemed a living creature, the crystals squirming. Through the narrow window of his microscope lens he saw the entire history of the earth. He witnessed the cosmos. He was no longer able to look away from the lens, and in his rapture he followed the dizzying creation and ruin of the crystals. His heart throbbed at the thought that he had glimpsed a universe whose true shape seldom revealed itself.

When at last the pain in his eyes became too much and he raised them from the microscope, his familiar storehouse attic would reappear before him, but its walls, desks, ceiling, the entire room seemed wondrously strange—solid, yet decidedly uncrystalline. At one moment he thought he saw a world in which everything had assumed the form of crystals, and now it seemed to him as if that world were swiftly fading, withdrawing. At once fearful and ecstatic, Manase surrendered once more to the magic and fixed his eyes upon the lens again.

When he heard the crowing of the neighbor's rooster Manase realized it was almost dawn. He put away his tools, glanced with his bloodshot eyes at his study, climbed down the steep ladder, and returned to the main house. ■

David Leveson

FROM "The Innocence of Rock" in *A Sense of the Earth* (1972)

There is an innocence about rock. The pebble in my hand, the boulder I stand on, the cliff wall next to my face that shades me are quiet and deceiving. Air and water rush and echo against their surface, but they themselves say nothing. If in an excess of energy I hurl the pebble into the sea or pry the boulder loose and let it roll down a slope, I may feel I am the master. In determining their position, their movement, I gain a temporary illusion of power, even if secretly I know that in the long run I am less enduring than they.

That this is an illusion, however, is clear. The boulder on which I stand, the pebble that I release or hold—it is they, really, that hold me. Together with the mass from which they have been broken, the mountains and plateaus where they originated, they state the framework of my existence: permit it, define its possibilities, outline its limitations. In short, they comprise the earth, whose own beginning and growth lay in the careless accumulation of fragments such as these—cosmic pebbles and boulders swept from space some five billion years ago—and whose aggregate material musters the feeble gravity that clasps us to it and keeps us from spinning off in the void.

Is this essential function performed in innocence, unknowing and mindless? There is a passive quality to rock that makes us think so. It suffers with seeming indifference our attempts to alter it, carve it, tunnel it, heap it up in mounds, to shape it to our will. With luck it doesn't cave in, shift or otherwise assert its presence. It is just there, ultimately important, immediately subordinate.

There are, of course, great belts around the earth—about the margin of the Pacific and through much of the Mediterranean and central Asia—where the passivity of rock is suspect, where stillness and predictability are rightly judged as momentary pauses in an almost continual round of quivering, shaking and erupting. There, rock is the enemy as well as the giver of life. A hostility emanates from it that transcends indifference of occasional moments of idleness.

But places where rock is known as fickle and places where it wears the mantle of eternity are, even within human experience, interchangeable. How short is memory; how great is our need to trust. If the flicker of time was quickened and the centuries passed as seconds, we could not ever forget the unstable nature of this globe that we inhabit. The evident drifting of its continents, the rapid rise and fall of its mountains, the swift draining of its seas like giant tide pools from one area to another, would impress deeply upon us the precarious character of our existence—and the ultimate generosity of the earth. A single lapse, a second without tenderness, and all rock would be cleansed of its living debris.

If rock is primarily our foothold on life, it is also a reminder, a link—if we let it be—of our relationship to and the reality of the rest of the universe. No less than the stars, the rocks so casually around us are the stuff of the world: their siblings circle the sun, their godparents populate all space. When the pebble in my hand returns the pressure I give it, its coolness or warmth, its quiet weight are not really an indifference. They are what is, for what else exists, really, after the pebble I hold and the boulder on which I stand?

There is love, violence, the complexity of life. But life is inextricably intertwined with the substance at hand, materially and psychologically. The shape of civilization, beyond what it has to extract from the earth, is evolved in the shade of the cliff, the heat of the plain, its fiber and warp bent to the contour and sweep of the land around it. Even in the heart of the cities, brick is from clay, cement derived from the limy excesses of antediluvian seas. In moments of hesitation or despair such knowledge can be a measure of sanity, a route to reality. The "natural" world is the world, and the productions of man extensions of it—always—even when obscene or tortuously remote. It helps to remember.

The variety of rock—its colors, textures, topology, its contrasts with life, light and water—is a deep and endless source of regeneration: the roughness of a stone fireplace, a walk through the visceral winding of a canyon, the pebble in your mouth to keep away thirst, the rattle of shingles on a beach, chance granite that forms a curb or paving block, the sheet of marble that rises to the sky. There is no stinginess about rock! Sought out or revealed unexpectedly, apparent or disguised, it is the shape and substance of things—within whose folds we are born, on whose flanks we are passionate and follow our fate, to whose depths we eventually return. It is rock from which the seas were squeezed, the atmosphere expelled, the accident of whose mass keeps our molecules close rather than flying, wasted, into space. ■

Luci Tapahonso

FROM "Ode to the Land: The Diné Perspective" (1995)

When a Diné child is born, part of the birth ritual includes the burying of the umbilical cord outside the family home. This ensures that throughout life, she or he will always return home, that the child will always care for his or her parents and that the child will not wander aimlessly as an adult. Since the Diné emerged from Mother Earth, Nahasdzáán in northwest New Mexico, burying the baby's cord signifies the importance of Nahasdzáán in the child's life, as well as her spiritual role in Diné history. It symbolizes the child's metaphorical and symbolic link to the earth. There are other rituals that emphasize this important concept.

We are taught that our land, Diné Tah or Navajo Country, was specifically created for us. There are many stories, songs and prayers that focus on various aspects of this knowledge. The informal and sometimes highly ritualized instruction is integrated into daily life from early childhood. The role of Nahasdzáán is easily understood in most stories, yet is intensely complex in symbolic and sacred terms.

Stories associated with landscape are plentiful and are often told as traveling stories, bedtime or mealtime stories or as part of explaining various characteristics of certain people or places. To be part of such stories confers a kind of respect, affection and joy in the participants' presence. People of all ages never tire of "hané," the telling and sharing of stories. It is in this spirit that the following stories associated with specific places in Diné country are presented.

Dzilná'o oditii (Huerfano Mesa)

Dzilná'o oditii is the center mountain of Diné Tah, the doorway to the six sacred places. Many centuries ago, Changing Woman, the most beloved of deities, was raised atop this mesa by First Woman and First Man. She later reared her sons, the Twin Warriors, here. The Holy People lived at Dzilná'o oditii and here the foundations of Diné thinking, knowledge and way of life were established. The presence of these Holy People is powerful on Dzilná'o oditii; the logs from the *hooghans* (hogans) remain, the cistern where Changing Woman bathed her babies remains and the footprints of the giant whom the Twins eventually slew are embedded atop the mesa.

This is considered the "doorway" to the past and to the future. As they created Dzilná'o oditii, the Holy People clothed her in precious fabrics as a symbol of soft goods. Thus, clothes of shiny, soft fabrics are especially valued among the Diné. "We dress as she does," we say. For special events, men and women, as well as the children, dress in velvet blouses and shirts, their silver and turquoise jewelry shines like clear water in a mountain stream. The dark, lush fabric lends an understated elegance to the men's black hats and the smooth, dark hair of the young girls and women. The long-tiered skirts flow soft and shiny. During annual fairs, parades, weddings, school programs and various events, the Diné put on their finest clothes. This is the attire considered to be the real "Navajo look." In doing this, we honor Changing Woman. In doing this, we dress as they have taught us. In doing this, we embody Dzilná'o oditii—the most sacred of places.

Rainbow over Chaco Canyon

In the old stories, the Diné say the Holy People traveled on sunbeams and a rainbow beam. Today, when a rainbow appears after a cleansing rain, it tells us the Holy Ones have returned. They remember and visit their children, those called the "Old Ones," the Anasazi at Chaco Canyon.

When they created this world, Blanca Peak, the sacred mountain in the east, was decorated with a rainbow beam and adorned with white shell and morning light. In the north, Mount Hesperus was fastened to the earth with a rainbow beam and adorned with black jet to represent peace and harmony. Each night, Mount Hesperus urges us to rest. She is our renewal, our rejuvenation. She exists because of the rainbow beam. We exist because of the rainbow beam.

When the Holy People return, they marvel at the growth of new spring plants, revel in the laughter of children splashing in fresh rain puddles and, like us, they inhale deeply the sweet clean air. A rainbow in the clear sky over Navajo land sparkles with particles of dew, pollen and the blessings of the Holy Ones.

Shiprock (Tsé Bit'a'í)

In one story about Tsé Bit'a'í, huge flying monsters once lived atop the Shiprock pinnacle and they would swoop down, snatching human prey to feed to their babies who remained in the nest. The people lived in fear. Finally, one of the Twin Warriors slew the birds, making it safe for the Diné to live here.

From the west, looking out from the Carriso Mountains, stands Little Shiprock. Shiprock looms in the background. My father was born almost 100 years ago, west of Little Shiprock, at Mitten Rock. This panorama reminds me of his childhood, our parents' first home, and the many relatives who live in the area around Oak Springs and Red Valley.

The vast, beautiful distance is filled with stories of people traveling on horseback across the distance, sometimes in wagons when children were taken to school, and sometimes in backs of pickup trucks filled with laundry and groceries. Always it fills one with a longing for drives that wind around sandstone formations, in and out of valleys, and the car filled with the silence of sleeping passengers, or with music—the radio playing or someone singing old Navajo songs. These drives are for telling stories—old stories of long ago, or maybe about something that happened yesterday. When one has grown up traveling miles and miles to school, work or for groceries, the hours spent driving become a time to share, to laugh, to teach and to simply be together.

I return often to my parents' home on the east side of Shiprock and, in the mornings, I walk through the furrowed fields with Chahbah, the family dog, and in the distance stands Shiprock. It is clothed in a soft, purple glow, majestic against the turquoise sky and surrounded by the dark blue mountain range in the west. As children, we woke to this sight each morning. Shiprock seemed to merge with the edges of our father's fields and so we thought it was part of our land. Each evening, the sunset's brilliant hues turn the huge rock into yellow, orange, red, pink, then purple before darkness moves in and it stands velvet black in the night.

In the darkness, it is easy to imagine huge monster birds perched atop the rock. It is natural to whisper thanks to the Twin Warriors and to Changing Woman for having raised them to be courageous and fearless.

Mount Taylor (Tsoodzit)

More than 100 years ago, as the Diné were returning from imprisonment at Fort Sumner, they wept at the first sight of Mount Taylor. "Now we'll surely make it home," they cried. They were weary from the long journey, but seeing Tsoodzit, the sacred mountain in the south, revived them physically and spiritually. They were strengthened because Tsoodzit represents adolescence—the strongest time of life. Tsoodzit helps us envision our goals and reminds us of our inner strength. Tsoodzit teaches us to believe in all ways of learning.

A stone knife was used to fasten Tsoodzit to the earth, then Tsoodzit was dressed in turquoise to represent the importance of positive thinking. It is the home of several Holy People, including Turquoise Girl and Turquoise Boy. Thus we wear turquoise in their honor, as well as to honor the males in our families.

Today, we tell the story over and over of our forebearers' return from Hweeldi and of all they endured and of the strength Tsoodzit gave them as they approached home.

Narbona Pass

When one travels through this pass, it is easy to imagine the huge sheer cliffs moving together to crush intruders as it was said to have done long ago. When the Twin Warriors went to visit their father, the Sun, they encountered many obstacles. With the help of Spider Woman, they were able to find a strong, unbreakable log to hold the huge rocks apart as they entered and passed through. Since then, the rocks have remained in place, the powerful danger dissipated by the Twin Warriors and the Holy People. The Twins continued on their journey and were eventually reunited with their father, who helped them rid the Earth of several deadly monsters who roamed about freely.

Centuries later, Narbona, an influential Diné leader, was killed in this pass by the United States Cavalry. Narbona had sought peace between the Diné and the U.S. forces, and had brought horses and sheep as part of the ongoing treaty negotiations. Because one of the Diné horses was mistakenly thought to have been stolen, Narbona was shot in the back four times. Six other Diné were also slain. Ironically, Narbona Pass was named "Washington Pass" after Col. John Washington, who had ordered the shooting. It was changed recently to honor Narbona.

Today, as people ski, picnic, hunt and hike in the area, they are appreciative of the overwhelming beauty, yet are aware of the historical significance Narbona Pass represents. They tell and retell the stories of the huge rocks and, in doing so, pay homage to the spirits of Narbona and the Twin Warriors.

Diné Tah/Petroglyphs at Diné Tah

At the beginning of Diné time, the Holy People dwelled in Diné Tah. They created the world from this powerful place. But they also hunted, cooked, slept, sang and raised children much as we do today. They constructed the first hooghans—the round-topped female one for everyday living and the ceremonial, forked-top hooghan. There was an abundance of game and plant life at Diné Tah. Fortified structures of mud and stone also were constructed to provide protection from enemy tribes. This is where the foundations of Diné life were established. Indeed, the patterns they set were intricate and complex, yet easy to incorporate into modern life.

Diné Tah is said to embody the beginning of Diné knowledge: our history, language, ceremonies and beliefs. Before they left to live within other sacred mountains, they instructed that drawings be made as a legacy for the present-day Diné. They ensured that the medicine people would have a source of ancient knowledge and it is further evidence of their concern for us.

There are symbols of renderings of ancient ritual masks and symbolic images of various Holy People. There are portrayals of corn, deer and antelope, all life sustaining elements. They range from the earliest of times and end at the time of Spanish and American contact in the mid-16th century.

Sadly, many of the drawings are marred by vandalism today. Nevertheless, as the Diné travel here to pray and sing, the quiet solitude and the powerful presence of the Holy Ones cannot be lessened by modern intrusions. The most sacred of places is made powerful by the history, stories, songs and prayers it contains. As we see this place, it is an experience of awe and gratitude. It is as if the Holy People are physically comforting us, encouraging us, smiling at us, strengthening us. That Diné Tah seems an empty, barren place suits us—we are among the most fortunate people in the world because of it. ■

Jacquetta Hawkes

FROM "Digression on Rocks, Soils and Men" in *A Land* (1951)

Life has grown from the rock and still rests upon it; because men have left it far behind, they are able consciously to turn back to it. We do turn back, for it has kept some hold over us. A liberal rationalist, Professor G. M. Trevelyan, can write of "the brotherly love that we feel . . . for trees, flowers, even for grass, nay even for rocks and water" and of "our brother the rock"; the stone of Scone is still used in the coronation of our kings.

The Church, itself founded on the rock of Peter, for centuries fought unsuccessfully against the worship of "sticks and stones." Such pagan notions have left memories in the circles and monoliths that still jut through the heather on our moorlands or stand naked above the turf of our downs. I believe that they linger, too, however faintly, in our church-yards—for who, even at the height of its popularity, ever willingly used cast-iron for a tombstone?

It is true that these stones were never simply themselves, but stood for dead men, were symbols of fertility, or, as at Stonehenge, were primarily architectural forms. But for worshippers the idea and its physical symbol are ambivalent; peasants worship the Mother of God and the painted doll in front of them; the peasants and herdsmen of prehistoric times honored the Great Mother or the Sky God, the local divinities or the spirits of their ancestors and also the stones associated with them. The Blue Stones of Stonehenge, for example, were evidently laden with sanctity. It seems that these slender monoliths were brought from Pembrokeshire to Salisbury Plain because in Wales they had already absorbed holiness from their use in some other sacred structure. There is no question here that the veneration must have been in part for the stones themselves.

Up and down the country, whether they have been set up by men, iso-lated by weathering, or by melting ice, conspicuous stones are commonly identified with human beings. Most of our Bronze Age circles and menhirs have been thought by the country people living round them to be men or women turned to stone. The names often help to express this identification and its implied sense of kinship; Long Meg and her Daughters, the Nine Maidens, the Bridestone and the Merry Maidens. It is right that they should most often be seen as women, for somewhere in the mind of everyone is an awareness of woman as earth, as rock, as matrix. In all these legends human beings have seen themselves melting back into rock, in their imaginations must have pictured the body, limbs and hair melting into smoke and solidifying into these blocks of sandstone, limestone and granite.

Some feeling that represents the converse of this idea arises from sculpture. I have never forgotten my own excitement on seeing in a Greek exhibition an unfinished statue in which the upper part of the body was perfect (though the head still carried a mantle of chaos) while the lower part disappeared into a rough block of stone. I felt that the limbs were already in existence, that the sculptor had merely been uncovering them, for his soundings were there—little tunnels reaching towards the position of the legs, feeling for them in the depths of the stone. The sculptor is in fact doing this, for the act of creation is in his mind, from his mind the form is projected into the heart of the stone, where then the chisel must reach it.

Rodin was one of the sculptors most conscious of these emotions, and most ready to exploit them. He expressed both aspects of the process— man merging back into the rock, and man detaching himself from it by the power of life and mind. He was perhaps inclined to sentimentalize the relationship by dwelling on the softness of the flesh in contrast with the rock's harshness. This was an irrelevance not dreamt of by the greatest exponent of the feeling—Michelangelo. It is fitting that the creator of the mighty figures of Night and Day should himself have spent many days in the marble quarries of Tuscany supervising the removal of his material from the side of the mountain. So conscious was he of the individual quality of the marble and of its influence on sculpture and architecture that he was willing to endure a long struggle with the Pope and at last to suffer heavy financial loss by maintaining the superiority of Carrara over Servezza marble. Michelangelo was an Italian working with Italian marble and Italian light; with us it has been unfortunate that since medieval times so many of our sculptors have sought the prestige of foreign stones rather than following the idiom of their native rock. It is part of the wisdom of our greatest sculptor, Henry Moore, to have returned to English stones and used them with a subtle sensitiveness for their personal qualities. He may have inherited something from his father who, as a miner spending his working life in the Carboniferous horizons of Yorkshire, must have had a direct understanding not only of coal but of the sandstones and shales in which it lies buried and on which the life of the miner depends. Henry Moore has himself made studies of miners at work showing their bodies very intimate with the rock yet charged with a life that separates them from it. (Graham Sutherland in his studies of tin mines became preoccupied with the hollow forms of the tunnels and in them his men appear almost embryonic.)

Henry Moore uses his understanding of the personality of stones in his sculpture, allowing their individual qualities to contribute to his conception. Indeed, he may for a moment be regarded in the passive role of a sympathetic agent giving expression to the stone, to the silting of ocean beds shown in those fine bands that curve with the sculpture's curves, and to the quality of the life that shows itself in the delicate markings made by shells, corals and sea-lilies.

It would certainly be inappropriate to his time if Moore habitually used the Italian marbles so much in favor since the Renaissance. For this fashion shows how man in his greatest pride of conscious isolation wanted stone which was no more than a beautiful material for his mastery. Now when our minds are recalling the past and our own origins deep within it, Moore reflects a greater humility in avoiding the white silence of marble and allowing his stone to speak. That is why he has often chosen a stone like Hornton, a rock from the Lias that is full of fossils all of which make their statement when exposed by his chisel. Sometimes the stone may be so assertive of its own qualities that he has to battle with it, strive against the hardness of its shells and the softness of adjacent pockets to make them, not efface themselves, but conform to his idea, his sense of a force thrusting from within, which must be expressed by taut lines without weakness of surface.

Moore uses Hornton stone also because it has two colors, a very pale brown and a green with deeper tones in it. The first serves him when he is conscious of his subject as a light one, the green when it must have darkness in it. Differences in climate round the shores of the Liassic lakes probably caused the change in color of Hornton stone, and so past climates are reflected in the feeling of these sculptures. As for the sculptor's sense of light or darkness inherent in his subjects, it is my belief that it derives in large part from the perpetual experience of day and night to which all consciousness has been subject since its beginnings. The sense of light and darkness seems to go to the depths of man's mind, and whether it is applied to morality, to aesthetics or to that more general conception—the light of intellectual processes in contrast with the darkness of the subconscious—its symbolism surely draws from our constant swing below the cone of night.

It is hardly possible to express in prose the extraordinary awareness of the unity of past and present, of mind and matter, of man and man's origin which these thoughts bring to me. Once when I was in Moore's studio and saw one of his reclining figures with the shaft of a belemnite exposed in the thigh, my vision of this unity was overwhelming. I felt that the squid in which life had created that shape, even while it still swam in distant seas was involved in this encounter with the sculptor; that it lay hardening in the mud until the time when consciousness was ready to find it out and imagination to incorporate it in a new form. So a poet will sometimes take fragments and echoes from other earlier poets to sink them in his own poems where they will enrich the new work as these fossil outlines of former lives enrich the sculptor's work.

Rodin pursued the idea of conscious, spiritual man emerging from the rock; Moore sees him rather as always a part of it. Through his visual similes he identifies women with caverns, caverns with eye-sockets; shells, bones, cell plasm drift into human form. Surely Mary Anning might have found one of his forms in the Blue Lias of Lyme Regis? That indeed would be fitting, for I have said that the Blue Lias is like the smoke of memory, of the subconscious, and Moore's creations float in those depths, where images melt into one another, the direct source of poetry, and the distant source of nourishment for the conscious intellect with its clear and fixed forms. I can see his rounded shapes like whales, his angular shapes like ichthyosaurs, surfacing for a moment into that world of intellectual clarity, but plunging down again to the sea bottom, the sea bottom where the rocks are silently forming.

Men know their affinity with rock and with soil, but they also use them, at first as simply as coral organisms use calcium, or as caddis-worms use shell and pebbles, but soon also consciously to express imagined ideas.

Building is one of the activities relating men most directly to their land. Everyone who travels inside Britain knows those sudden changes between region and region, from areas where houses are built of brick or of timber and daub and fields are hedged, to those where houses are of stone and fields enclosed by drystone walling. Everywhere in the ancient mountainous country of the west and north stone is taken for granted; where the sudden appearance of walls instead of hedges catches the eye is along the belt of Jurassic limestones, often sharply delimited. The change is most dramatic in Lincolnshire where the limestone of Lincoln Edge is not more than a few miles wide and the transformation from hedges to the geometrical austerity of dry-walling, from the black and white, red and buff of timber and brick to the melting greys of limestone buildings, is extraordinarily abrupt.

The distinctive qualities of the stones of each geological age and of each region powerfully affect the architecture raised up from them; if those qualities precisely meet particular needs then, of course, the stones are carried out of their own region. Since the eighteenth century the value of special qualities in building material has greatly outweighed the labor of transport, and stones of many kinds have not only been carried about Britain to places far from those where they were originally formed, but have been sent overseas to all parts of the world. Men, in fact, have proved immensely more energetic than rivers or glaciers in transporting and mixing the surface deposits of the planet. ∎

E. M. Forster

FROM *A Passage to India* (1924)

The Ganges, though flowing from the foot of Vishnu and through Siva's hair, is not an ancient stream. Geology, looking further than religion, knows of a time when neither the river nor the Himalayas that nourished it existed, and an ocean flowed over the holy places of Hindustan. The mountains rose, their debris silted up the ocean, the gods took their seats on them and contrived the river, and the India we call immemorial came into being. But India is really far older. In the days of the prehistoric ocean the southern part of the peninsula already existed, and the high places of Dravidia have been land since land began, and have seen on the one side the sinking of a continent that joined them to Africa, and on the other the upheaval of the Himalayas from a sea. They are older than anything in the world. No water has ever covered them, and the sun who has watched them for countless aeons may still discern in their outlines forms that were his before our globe was torn from his bosom. If flesh of the sun's flesh is to be touched anywhere, it is here, among the incredible antiquity of these hills.

Yet even they are altering. As Himalayan India rose, this India, the primal, has been depressed, and is slowly re-entering the curve of the earth. It may be that in eons to come an ocean will flow here too, and cover the sun-born rocks with slime. Meanwhile the plain of the Ganges encroaches on them with something of the sea's action. They are sinking beneath the newer lands. Their main mass is untouched, but at the edge their outposts have been cut off and stand knee-deep, throat-deep, in the advancing soil. There is something unspeakable in these outposts. They are like nothing else in the world, and a glimpse of them makes the breath catch. They rise abruptly, insanely, without the proportion that is kept by the wildest hills elsewhere, they bear no relation to anything dreamt or seen. To call them "uncanny" suggests ghosts, and they are older than all spirit. Hinduism has scratched and plastered a few rocks, but the shrines are unfrequented, as if pilgrims, who generally seek the extraordinary, had here found too much of it. Some saddhus did once settle in a cave, but they were smoked out, and even Buddha, who must have passed this way down to the Bo Tree of Gya, shunned a renunciation more complete than his own, and has left no legend of struggle or victory in the Marabar.

The caves are readily described. A tunnel eight feet long, five feet high, three feet wide, leads to a circular chamber about twenty feet in diameter. This arrangement occurs again and again throughout the group of hills, and this is all, this is a Marabar Cave. Having seen one such cave, having seen two, having seen three, four, fourteen, twenty-four, the visitor returns to Chandrapore uncertain whether he has had an interesting experience or a dull one or any experience at all. He finds it difficult to discuss the caves, or to keep them apart in his mind, for the pattern never varies, and no carving, not even a bees'-nest or a bat distinguishes one from another. Nothing, nothing attaches to them, and their reputation—for they have one—does not depend on human speech. It is as if the surrounding plain or the passing birds have taken upon themselves to exclaim "extraordinary," and the world has taken root into the air, and been inhaled by mankind.

They are dark caves. Even when they open towards the sun, very little light penetrates down the entrance tunnel into the circular chamber. There is little to see, and no eye to see it, until the visitor arrives for his five minutes, and strikes a match. Immediately another flame rises in the depths of the rock and moves towards the surface like an imprisoned spirit: the walls of the circular chamber have been most marvelously polished. The two flames approach and strive to unite, but cannot, because one of them breathes air, the other stone. A mirror inlaid with lovely colors divides the lovers, delicate stars of pink and grey interpose, exquisite nebulae, shadings fainter than the tail of a comet or the midday moon, all the evanescent life of the granite, only here visible. Fists and fingers thrust above the advancing soil—here at last is their skin, finer than any covering acquired by the animals, smoother than windless water, more voluptuous than love. The radiance increases, the flames touch one another, kiss, expire. The cave is dark again, like all the caves.

Only the wall of the circular chamber has been polished thus. The sides of the tunnel are left rough, they impinge as an afterthought upon the internal perfection. An entrance was necessary, so mankind made one. But elsewhere, deeper in the granite, are there certain chambers that have no entrances? Chambers never unsealed since the arrival of the gods. Local report declares that these exceed in number those that can be visited, as the dead exceed the living—four hundred of them, four thousand or million. Nothing is inside of them, they were sealed up before the creation of pestilence or treasure; if mankind grew curious and excavated, nothing, nothing would be added to the sum of good or evil. One of them is rumored within the boulder that swings on the summit of the highest of the hills; a bubble-shaped cave that has neither ceiling nor floor, and mirrors its own darkness in every direction infinitely. If the boulder falls and smashes, the cave will smash too—empty as an Easter egg. The boulder because of its hollowness sways in the wind, and even moves when a crow perches upon it: hence its name and the name of its stupendous pedestal: the Kawa Dol. ▪

Loren Eiseley

FROM "The Relic Men" in *The Night Country* (1971)

I remember the sound of the wind in that country never stopped. I think everyone there was a little mad because of it. In the end I suppose I was like all the rest. It was a country of topsyturvy, where great dunes of sand blew slowly over ranch houses and swallowed them, and where, after the sand had all blown away from under your feet, the beautiful arrowheads of ice-age hunters lay mingled with old whisky bottles that the sun had worked upon. I suppose, now that I stop to think about it, that if there is any place in the world where a man might fall in love with a petrified woman, that may be the place.

In the proper books, you understand, there is no such thing as a petrified woman, and I insist that when I first came to that place I would have said the same. It all happened because bone hunters are listeners. They have to be.

We had had terrible luck that season. We had made queries in a score of towns and tramped as many canyons. The institution for which we worked had received a total of one Oligocene turtle and a bag of rhinoceros bones. A rag picker could have done better. The luck had to change. Somewhere there had to be fossils.

I was cogitating on the problem under a coating of lather in a barbershop with an 1890 chair when I became aware of a voice. You can hear a lot of odd conversation in barbershops, particularly in the back country, and particularly if your trade makes you a listener, as mine does. But what caught my ear at first was something about stone. Stone and bone are pretty close in my language and I wasn't missing any bets. There was always a chance that there might be a bone in it somewhere for me.

The voice went off into a grumbling rural complaint in the back corner of the shop, and then it rose higher.

"It's petrified! It's petrified!" the voice contended excitedly.

I managed to push an ear up through the lather.

"I'm a-tellin' ya," the man boomed, "a petrified woman, right out in that canyon. But he won't show it, not to nobody. 'Tain't fair, I tell ya."

"Mister," I said, speaking warily between the barber's razor and his thumb, "I'm reckoned a kind of specialist in these matters. Where is this woman, and how do you know she's petrified?"

I knew perfectly well she wasn't, of course. Flesh doesn't petrify like wood or bone, but there are plenty of people who think it does. In the course of my life I've been offered objects that ranged from petrified butterflies to a gentleman's top hat.

Just the same, I was still interested in this woman. You can never tell what will turn up in the back country. Once I had a mammoth vertebra handed to me with the explanation that it was a petrified griddle cake. Mentally, now, I was trying to shape that woman's figure into the likeness of a mastodon's femur. This is a hard thing to do when you are young and far from the cities. Nevertheless, I managed it. I held that shining bony vision in my head and asked directions of my friend in the barbershop.

Yes, he told me, the woman was petrified all right. Old man Buzby wasn't a feller to say it if it 'tweren't so. And it weren't no part of a woman. It was a *whole* woman. Buzby had said that, too. But Buzby was a queer one. An old bachelor, you know. And when the boys had wanted to see it, 'count of it bein' a sort of marvel around these parts, the old man had clammed up on where it was. A-keepin' it all to hisself, he was. But seein' as I was interested in these things and a stranger, he might talk to me and no harm done. It was the trail to the right and out and up to the overhang of the hills. A little tarpapered shack there.

I asked Mack to go up there with me. He was silent company but one of the best bone hunters we had. Whether it was a rodent the size of a bee or an elephant the size of a house, he'd find it and he'd get it out, even if it meant that we carried a five-hundred-pound plaster cast on foot over a mountain range.

In a day we reached the place. When I got out of the car I knew the wind had been blowing there since time began. There was a rusty pump in the yard and rusty wire and rusty machines nestled in the lee of a wind-carved butte. Everything was leaching and blowing away by degrees, even the tarpaper on the roof.

Out of the door came Buzby. He was not blowing away, I thought at first. His farm might be, but he wasn't. There was an air of faded dignity about him.

Now in that country there is a sort of etiquette. You don't drive out to a man's place, a bachelor's, and you a stranger, and come up to his door and say: "I heard in town you got a petrified woman here, and brother, I sure would like to see it." You've got to use tact, same as anywhere else.

You get out slowly while the starved hounds look you over and get their barking done. You fumble for your pipe and explain casually you're doin' a little lookin' around in the hills. About that time they get a glimpse of the equipment you're carrying and most of them jump to the conclusion that you're scouting for oil. You can see the hope flame up in their eyes and sink down again as you explain you're just hunting bones. Some of them don't believe you after that. It's a hard thing to murder a poor man's dream.

But Buzby wasn't the type. I don't think he even thought of the oil. He was small and neat and wore—I swear it—pince-nez glasses. I could see at a glance that he was a city man dropped, like a seed, by the wind. He had been there a long time, certainly. He knew the corn talk and the heat talk, but he would never learn how to come forward in that secure, heavy-shouldered country way, to lean on a car door and talk to strangers while the horizon stayed in his eyes.

He invited us, instead, to see his collection of arrowheads. It looked like a good start. We dusted ourselves and followed him in. It was a two-room shack, and about as comfortable as a monk's cell. It was neat, though, so neat you knew that the man lived, rather than slept there. It lacked the hound-asleep-in-the-bunk confusion of the usual back-country bachelor's quarters.

He was precise about his Indian relics as he was precise about every-thing, but I sensed after a while a touch of pathos—the pathos of a man clinging to order in a world where the wind changed the landscape before morning, and not even a dog could help you contain the loneliness of your days.

"Someone told me in town you might have a wonderful fossil up here," I finally ventured, poking in his box of arrowheads, and watching the shy, tense face behind the glasses.

"That would be Ned Burner," he said. "He talks too much."

"I'd like to see it," I said, carefully avoiding the word *woman*. "It might be something of great value to science."

He flushed angrily. In the pause I could hear the wind beating at the tarpaper.

"I don't want any of 'em hereabouts to see it," he cried passionately. "They'll laugh and they'll break it and it'll be gone like—like everything." He stopped, uncertainly aware of his own violence, his dark eyes widening with pain.

"We are scientists, Mr. Buzby," I urged gently. "We're not here to break anything. We don't have to tell Ned Burner what we see."

He seemed a little mollified at this, then a doubt struck him. "But you'd want to take her away, put her in a museum."

I noticed the pronoun but ignored it. "Mr. Buzby," I said, "we would very much like to see your discovery. It may be we can tell you more about it that you'd like to know. It might be that a museum would help you save it from vandals. I'll leave it to you. If you say no, we won't touch it, and we won't talk about it in the town, either. That's fair enough, isn't it?"

I could see him hesitating. It was plain that he wanted to show us, but the prospect was half-frightening. Oddly enough, I had the feeling his fright revolved around his discovery, more than fear of the townspeople. As he talked on, I began to see what he wanted. He intended to show it to us in the hope we would confirm his belief that it was a petrified woman. The whole thing seemed to have taken on a tremendous importance in his mind. At that point, I couldn't fathom his reasons.

Anyhow, he had something. At the back of the house we found the skull of a big, long-horned, extinct bison hung up under the eaves. It was a nice find, and we coveted it.

"It needs a dose of alvar for preservation," I said. "The museum would be the place for a fine specimen like this. It will just go slowly to pieces here."

Buzby was not unattentive. "Maybe, Doctor, maybe. But I have to think. Why don't you camp here tonight? In the morning—"

"Yes?" I said, trying to keep the eagerness out of my voice. "You think we might—?"

"No! Well, yes, all right. But the conditions? They're like you said?"

"Certainly," I answered. "It's very kind of you."

He hardly heard me. That glaze of pain passed over his face once more. He turned and went into the house without speaking. We did not see him again until morning.

The wind goes down into those canyons also. It starts on the flats and rises through them with weird noises, flaking and blasting at every loose stone or leaning pinnacle. It scrapes the sand away from pipy concretions till they stand out like strange distorted sculptures. It leaves great stones teetering on wineglass stems.

I began to suspect what we would find, the moment I came there. Buzby hurried on ahead now, eager and panting. Once he had given his consent and started, he seemed in almost frenzy of haste.

Well, it was the usual thing. Up. Down. Up. Over boulders and splintered deadfalls of timber. Higher and higher into the back country. Toward the last he outran us, and I couldn't hear what he was saying. The wind whipped it away.

But there he stood, finally, at a niche under the canyon wall. He had his hat off and, for a moment, was oblivious to us. He might almost have been praying. Anyhow I stood back and waited for Mack to catch up. "This must be it," I said to him. "Watch yourself." Then we stepped forward.

It was a concretion, of course—an oddly shaped lump of mineral matter—just as I had figured after seeing the wind at work in those miles of canyon. It wasn't a bad job, at that. There were some bumps in the right places, and a few marks that might be the face, if your imagination was strong. Mine wasn't just then. I had spent a day building a petrified woman into a mastodon femur, and now that was no good either, so I just stood and looked.

But after the first glance it was Buzby I watched. The unskilled eye can build marvels of form where the educated see nothing. I thought of that bison skull under his eaves, and how badly we needed it.

He didn't wait for me to speak. He blurted with a terrible intensity that embarrassed me, "She—she's beautiful, isn't she?"

"It's remarkable," I said. "Quite remarkable." And then I just stood there not knowing what to do.

He seized on my words with such painful hope that Mack backed off and started looking for fossils in places where he knew perfectly well there weren't any.

I didn't catch it all; I couldn't possibly. The words came out in a long, aching torrent, the torrent dammed up for years in the heart of a man not meant for this place, nor for the wind at night by the windows, nor the empty bed, nor the neighbors twenty miles away. You're tough at first. He must have been to stick there. And then suddenly you're old. You're old and you're beaten, and there must be something to talk to and to love. And if you haven't got it you'll make it in your head, or out of a stone in a canyon wall.

He had found her, and he had a myth of how she came there, and now he came up and talked to her in the long afternoon heat while the dust devils danced in his failing corn. It was progressive. I saw the symptoms. In another year, she would be talking to him.

"It's true, isn't it, Doctor?" he asked me, looking up with that rapt face, after kneeling beside the niche. "You can see it's her. You can see it plain as day." For the life of me I couldn't see anything except a red scar writhing on the brain of a living man who must have loved somebody once, beyond words and reason.

"Now Mr. Buzby," I started to say then, and Mack came up and looked at me. This, in general, is when you launch into careful explanation of how concretions are made so that the layman will not make the same mistake again. Mack just stood there looking at me in that stolid way of his. I couldn't go on with it. I couldn't even say it.

But I saw where this was going to end. I saw it suddenly and too late. I opened my mouth while Mr. Buzby held his hands and tried to regain his composure. I opened my mouth and I lied in a way to damn me forever in the halls of science.

I lied, looking across at Mack, and I could feel myself getting redder every moment. It was a stupendous, a colossal lie. "Mr. Buzby," I said, "that—um—er—figure is astonishing. It is a remarkable case of preservation. We must have it for the museum."

The light in his face was beautiful. He believed me now. He believed himself. He came up to the niche again, and touched her lovingly.

"It's okay," I whispered to Mack. "We won't have to pack the thing out. He'll never give her up."

That's where I was a fool. He came up to me, his eyes troubled and unsure, but very patient.

"I think you're right, Doctor," he said. "It's selfish of me. She'll be safer with you. If she stays here somebody will smash her. I'm not well." He sat down on a rock and wiped his forehead. "I'm sure I'm not well. I'm sure she'll be safer with you. Only I don't want her in a glass case where people can stare at her. If you can promise that, I—"

"I can promise that," I said, meeting Mack's eyes across Buzby's shoulder.

"And if I come there I can see her!"

I knew I would never meet him again in this life.

"Yes," I said, "you can see her there." I waited, and then I said, "We'll get the picks and plaster ready. Now that bison skull at your house . . ."

It was two days later, in the truck, that Mack spoke to me. "Doc."

"Yeah."

"You know what the Old Man is going to say about shipping that concretion. It's heavy. Must be three hundred pounds with the plaster."

"Yes, I know."

Mack was pulling up slow along the abutment of a bridge. It was the canyon of the big Piney, a hundred miles away. He got out and went to the rear of the truck. I didn't say anything, but I followed him back.

"Doc, give me a hand with this, will you?"

I took one end, and we heaved together. It's a long drop in the big Piney. I didn't look, but I heard it break on the stones.

"I wish I hadn't done that," I said.

"It was only a concretion," Mack answered. "The old geezer won't know."

"I don't like it," I said. "Another week in that wind and I'd have believed in her myself. Get me the hell out of here—maybe I do, anyhow. I tell you I don't like it. I don't like it at all."

"It's a hundred more to Valentine," Mack said.

He put the map in the car pocket and slid over and gave me the wheel. ▪

Ann Zwinger

"A Question of Conglomerate" (previously unpublished)

The drive between Colorado Springs, where I live, and Denver, where I frequently attend meetings, is a beautiful seventy miles, visually and geologically. As I drive north, the short-grass prairie gleams in the morning's raking light, rolling a thousand miles eastward and six thousand feet downward to the Mississippi; to the west the land sweeps upward to the crest of the Rocky Mountains at fourteen thousand feet, mountains that crumpled upward during a period of intense mountain building that ended around 40 million years ago. The Continental Divide appears on the western horizon as a stunning snow-white ruching against a brilliant blue Colorado sky. I drive suspended between mountain and prairie, two disparate worlds unified into a landscape of vast majesty.

A sign that reads 7,325 feet marks the crest of Monument Hill, and the divide between the drainages of the Arkansas River to the south and the Platte River tributaries to the north. From here the road slopes down toward Denver, crossing a basin that sank as the Rocky Mountains rose. Huge sandstone pillars and castles, set in the forested foothills, stand out like monumental chess pieces, pale yellowish sandstone bright against the dark conifers. Even as these sandstones formed at the edge of old rivers and seas, water ate away the grains until eventually only these particular columns remained standing because they were both more strongly cemented and protected beneath younger patches of much harder rock that resisted erosion.

The right front seat asks, "What on earth are those rocks out there?" Then adds, somewhat wistfully, "I wish I knew more about geology. But it's just too complicated!"

Sigh. As a natural history writer, I hear this again and again from readers who complain about geologic descriptions in the text: geology is too difficult, too complicated, too hard to understand. No use explaining that to truly enjoy a landscape means finding out how and when it got there, what underlies it, what crops out of it, where did the winds come from, and the water—was it some ancient sea that lapped here, whose silts preserved delicate fossils, or a river sluicing off the melting glaciers of a past Ice Age? Don't misunderstand me: I have a lot of sympathy for my readers. Since I've never had a geology course, it's often been trying work to understand a concept well enough to try to explain it to a reader. But in the doing, I've come to appreciate (if not downright envy) the geologist's world.

Much of the resistance to matters geological, I think, must partly lie in our inability to take a long-term view, the *really* long-term view. We tend to judge events by our life spans, in weeks or months or years or maybe centuries. Geology requires a different way of looking at time. We can watch a building go up or a highway built, but layers of sediments that require ten thousand years to produce a half-inch layer are difficult to comprehend. It's a vast disconnect: I'm thinking about getting to Denver in sixty-five minutes while these sandstone columns are on the order of 65 million years old.

Ahead lies an unmistakable landmark, Castle Rock, the mesa that gives the town at its feet its name. It's the "type section" for which the Castle Rock Conglomerate is named, the place where this formation is most clearly revealed and from which it was originally described. It formed as an alluvial fan when cobbles, pebbles, and larger rocks washed down from the highlands to the west. Over time it indurated into a very hard rock, glued together by sand and silt, that now protects the softer rocks beneath and forms this imposing mesa.

A few years ago I too wondered what these same rock outcrops were and, for once, when I got home, looked up "Castle Rock" in *Roadside Geology of Colorado* (almost every state has one of these wonderful guides). I admit that it makes my day to know where I am and what I'm seeing, tuning into a world so long ago and different from my own.

The right front seat works on envisioning the forces that combined to create this landscape that pleases and puzzles us in the here and now. Maybe just being observant while in the actual context of these magnificent surroundings encourages us to be curious about an intricate, interactive, ever-changing past and urges us to ask the "how and why" questions. As for me, getting in touch with the geology that surrounds me provides entrée into a larger, reasonably working world that is very reassuring in unreassuring times, and rife with more intriguing questions to keep me out of mischief than I'll ever have time to find answers for. ▪

Jane Hirshfield

FROM *Given Sugar, Given Salt* (2001)

ROCK

What appears to be stubbornness,
refusal, or interruption,
is to it a simple privacy. It broods
its one thought like a quail her clutch of eggs.

Mosses and lichens
listen outside the locked door.
Stars turn the length of one winter, then the next.

Rocks fill their own shadows without hesitation,
and do not question silence,
however long.
Nor are they discomforted by cold, by rain, by heat.

The work of a rock is to ponder whatever is:
an act that looks singly like prayer,
but is not prayer.

As for this boulder,
its meditations are slow but complete.

Someday, its thinking worn out, it will be
carried away by an ant.
A *Mystrium camillae,*
perhaps, caught in some equally diligent,
equally single pursuit of a thought of her own. ▪

2 Deep Time

It's no metaphor to feel the influence of the dead in
the world, just as it's no metaphor to hear the radio-
carbon chronometer, the Geiger counter amplifying
the faint breathing of rock, fifty thousand years old.
(Like the faint thump from behind the womb wall.)
It's no metaphor to witness the astonishing fidelity
of minerals magnetized, even after hundreds of
millions of years, pointing to the magnetic pole,
minerals that have never forgotten magma whose
cooling off has left them forever desirous. We long
for place; but place itself longs. Human memory is
encoded in air currents and river sediment.

—Anne Michaels, *Fugitive Pieces*

HAVE YOU EVER TRIED TO IMAGINE ETERNITY? HUMAN PERCEPTIONS OF time and Earth's antiquity have varied through the ages and across cultures. Conceived of as infinite and finite, linear and circular, continuous and fragmented, time as a vector has stood in human thought as metaphor, myth, and scientific "absolute." But how do we comprehend the immensity of all history, of what American writer John McPhee has called "deep time"?

If we consider the geologic workings of Earth, erosion and tectonic forces have constantly reworked and recycled its landscapes through time, leaving records that are, at best, incompletely preserved. The land consists of accumulated and inherited pieces from different pasts of process and response. The constant dynamics are change, fragmentation, and assembly.

Thus the materials of any landscape—rocks and pebbles, sediment as strata, and landforms (both external shape and internal structure)—contain only partial records of former worlds, of environmental change, and, in the past several millennia, of human activities. Scientific studies of these features involve extensive field analysis of Earth's surface features and the use of sophisticated instruments to measure the composition or age of rocks, the shape of rock bodies at depth within the Earth, and the nature of the planet's inaccessible interior. Because of the complexity of Earth's fragmented history, the practice of geology is akin to detective work—a forensic effort to piece together the deep past from available evidence. The missing record means that much of the story is unknown and unknowable.

This section contains writings from the past three centuries that attempt to come to grips with the immense abyss of "deep time" and recognize the landscapes we inhabit as artifacts of an ancient and changing world, as ephemeral features, and as elements of the future.

The section opens with an excerpt from *Portrait of the Artist as a Young Man* (1916), in which Irish author James Joyce (1882–1941) imagines the "awful meaning" of immeasurable eternity. In the essay "The Slit" from *The Immense Journey* (1957), anthropologist and naturalist Loren Eiseley (1907–77) describes his encounter with a fossil skull exposed in a deep crack cut into a treeless plain. This experience leads him to contemplate the immensity of the time journey we all are engaged in and the small part that each of our individual experiences makes of the whole.

Antoine de Saint-Exupéry (1900–1944) was a pioneer of aviation and one of the great twentieth-century French writers. This author of *The Little Prince* died on an Allied mission in 1944, when his plane went down over the Mediterranean Sea. From his reflective *Wind, Sand and Stars* (1939) we include an account of a forced landing in the Sahara Desert.

The poem "The Fish in the Stone" by contemporary poet Rita Dove (1952–) follows. Dove is the author of several books of poetry, including *Thomas and Beulah* (1986), which won the Pulitzer Prize for Poetry. She served as Poet Laureate and Consultant in Poetry to the Library of Congress from 1993 to 1995. She was the youngest person—and the first African-American woman—to receive this highest official honor in American letters.

From Edinburgh, Scotland, James Hutton (1726–97) was educated as a physician, but he concerned himself primarily with agriculture and study of the Earth. With the publication of his *Theory of the Earth*, study of the formation and history of the planet gained a scientific basis. Considered the founder of "modern" geology as a science, Hutton observed Earth processes and materials as key to deciphering the structure, composition, and history of the planet. The excerpt from the 1795 version of *Theory of the Earth* considers "the surface of this land . . . made by nature to decay" and the indefinite realm of time with "no vestige of a beginning—no prospect of an end."

Perceptions and narratives of the vast erosional badland topography of the semi-arid high western plains of North America, beyond the hundredth meridian, are varied and numerous. In *Man in the Landscape* (1967), human ecologist and philosopher Paul Shepard (1925–96) considers responses of westering overlanders to "novel" erosional remnants and angular cliffs and escarpments along the Oregon Trail in 1840s. As described in the excerpt, the isolated spires, buttes, and escarpments along the North Platte River in the semi-arid western plains became structures of ghostly or ruined architecture in journals of trappers, soldiers, and emigrants. Observers discovered city buildings, lighthouses, forts, spires, streets, and castles, often in ruins; and many references were made to rocks given names like Chimney, Steamboat, Table, or Courthouse.

We end this section with Mark Twain's satirical essay "Was the World Made for Man?" from "The Damned Human Race" in the collection *Letters from the Earth,* edited by Bernard DeVoto.

James Joyce

FROM *Portrait of the Artist as a Young Man* (1916)

You have often seen the sand on the seashore. How fine are its tiny grains! And how many of those tiny little grains go to make up the small handful which a child grasps in its play. Now imagine a mountain of that sand, a million miles high, reaching from the earth to the farthest heavens, and a million miles broad, extending to remotest space, and a million miles in thickness: and imagine such an enormous mass of countless particles of sand multiplied as often as there are leaves in the forest, drops of water in the mighty ocean, feathers on birds, scales on fish, hairs on animals, atoms in the vast expanse of the air: and imagine that at the end of every million years a little bird came to that mountain and carried away in its beak a tiny grain of that sand. How many millions upon millions of centuries would pass before that bird had carried away even a square foot of that mountain, how many eons upon eons of ages before it had carried away all. Yet at the end of that immense stretch of time not even one instant of eternity could be said to have ended. At the end of all those billions and trillions of years eternity would have scarcely begun. And if that mountain rose again after it had been all carried away and if the bird came again and carried it all away again grain by grain: and if it so rose and sank as many times as there are stars in the sky, atoms in the air, drops of water in the sea, leaves on the trees, feathers upon birds, scales upon fish, hairs upon animals, at the end of all those innumerable risings and sinkings of that immeasurably vast mountain not one single instant of eternity could be said to have ended; even then, at the end of such a period, after that eon of time the mere thought of which makes our very brain reel dizzily, eternity would scarcely have begun. ▪

Loren Eiseley

"The Slit" in *The Immense Journey* (1957)

Some lands are flat and grass-covered, and smile so evenly up at the sun that they seem forever youthful, untouched by man or time. Some are torn, ravaged and convulsed like the features of profane old age. Rocks are wrenched up and exposed to view; black pits receive the sun but give back no light.

It was to such a land I rode, but I rode to it across a sunlit, timeless prairie over which nothing passed but antelope or a wandering bird. On the verge where that prairie halted before a great wall of naked sandstone and clay, I came upon the Slit. A narrow crack worn by some descending torrent had begun secretly, far back in the prairie grass, and worked itself deeper and deeper into the fine sandstone that led by devious channels into the broken waste beyond. I rode back along the crack to a spot where I could descend into it, dismounted, and left my horse to graze.

The crack was only about body-width and, as I worked my way downward, the light turned dark and green from the overhanging grass. Above me the sky became a narrow slit of distant blue, and the sandstone was cool to my hands on either side. The Slit was a little sinister—like an open grave, assuming the dead were enabled to take one last look—for over me the sky seemed already as far off as some future century I would never see.

I ignored the sky, then, and began to concentrate on the sandstone walls that had led me into this place. It was tight and tricky work, but that cut was a perfect cross section through perhaps ten million years of time. I hoped to find at least a bone, but I was not quite prepared for the sight I finally came upon. Staring straight out at me, as I slid farther and deeper into the green twilight, was a skull embedded in the solid sandstone. I had come at just the proper moment when it was fully to be seen, the white bone gleaming there in a kind of ashen splendor, water worn, and about to be ground away in the next long torrent.

It was not, of course, human. I was deep, deep below the time of man in a remote age near the beginning of the reign of mammals. I squatted on my heels in the narrow ravine, and we stared a little blankly at each other, the skull and I. There were marks of generalized primitiveness in that low, pinched brain case and grinning jaw that marked it as lying far back along those converging roads where, as I shall have occasion to establish elsewhere, cat and man and weasel must leap into a single shape.

It was the face of a creature who had spent his days following his nose, who was led by instinct rather than memory, and whose power of choice was very small. Though he was not a man, nor a direct human ancestor, there was yet about him, even in the bone, some trace of that low, snuffling world out of which our forebears had so recently emerged. The skull lay tilted in such a manner that it stared, sightless, up at me as though I, too, were already caught a few feet above him in the strata and, in my turn, were staring upward at that strip of sky which the ages were carrying farther away from me beneath the tumbling debris of falling mountains. The creature had never lived to see a man, and I, what was it I was never going to see?

I restrained a panicky impulse to hurry upward after that receding sky that was outlined above the Slit. Probably, I thought, as I patiently began the task of chiseling into the stone around the skull, I would never again excavate a fossil under conditions which led to so vivid an impression that I was already one myself. The truth is that we are all potential fossils still carrying within our bodies the crudities of former existences, the marks of a world in which living creatures flow with little more consistency than clouds from age to age.

As I tapped and chiseled there in the foundations of the world, I had ample time to consider the cunning manipulability of the human fingers. Experimentally I crooked one of the long slender bones. It might have been silica, I thought, or aluminum, or iron—the cells would have made it possible. But no, it is calcium, carbonate of lime. Why? Only because of its history. Elements more numerous than calcium in the earth's crust could have been used to build the skeleton. Our history is the reason—we came from the water. It was there the cells took the lime habit, and they kept it after we came ashore.

It is not a bad symbol of that long wandering, I thought again—the human hand that has been fin and scaly reptile foot and furry paw. If a stone should fall (I cocked an eye at the leaning shelf above my head and waited, fatalistically) let the bones lie here with their message, for those who might decipher it, if they come down late among us from the stars.

Above me the great crack seemed to lengthen.

Perhaps there is no meaning in it at all, the thought went on inside me, save that of journey itself, so far as men can see. It has altered with the chances of life, and the chances brought us here; but it was a good journey—long, perhaps—but a good journey under a pleasant sun. Do not look for a purpose. Think of the way we came and be a little proud. Think of this hand—the utter pain of its first venture on the pebbly shore.

Or consider its later wanderings.

I ceased my tappings around the sand-filled sockets on the skull and wedged myself into a crevice for a smoke. As I tamped a load of tobacco into my pipe, I thought of a town across the valley that I used sometimes to visit, a town whose little inhabitants never welcomed me. No sign points to it and I rarely go there any more. Few people know about it and fewer still know that in a sense we, or rather some of the creatures to whom we are related, were driven out of it once, long ago. I used to park my car on a hill and sit silently observant, listening to the talk ringing out from neighbor to neighbor, seeing the inhabitants drowsing in their doorways, taking it all in with nostalgia—the sage smell on the wind, the sunlight without time, the village without destiny. We can look, but we can never go back. It is prairie-dog town.

"Whirl is king," said Aristophanes, and never since life began was Whirl more truly king than eighty million years ago in the dawn of the Age of Mammals. It would come as a shock to those who believe firmly that the scroll of the future is fixed and the road determined in advance, to observe the teetering balance of earth's history through the age of the Paleocene. The passing of the reptiles had left a hundred uninhabited life zones and a scrambling variety of newly radiating forms. Unheard-of species of giant ground birds threatened for a moment to dominate the earthly scene. Two separate orders of life contended at slightly different intervals for the pleasant grasslands—for the seed and the sleepy burrow in the sun.

Sometimes, sitting there in the mountain sunshine above prairie-dog town, I could imagine the attraction of that open world after the fern forest damp or the croaking gloom of carboniferous swamps. There by a tree root I could almost make him out, that shabby little Paleocene rat, eternal tramp and world wanderer, father of all mankind. He ruffed his coat in the sun and hopped forward for a seed. It was to be a long time before he would be seen on the grass again, but he was trying to make up his mind. For good or ill there was to be one more chance, but that chance was fifty million years away.

Here in the Paleocene occurred the first great radiation of the placental mammals, and among them were the earliest primates—the zoological order to which man himself belongs. Today, with a few unimportant exceptions, the primates are all arboreal in habit except man. For this reason we have tended to visualize all of our remote relatives as tree dwellers. Recent discoveries, however, have begun to alter this one-sided picture. Before the rise of the true rodents, the highly successful order to which present-day prairie dogs and chipmunks belong, the environment which they occupy had remained peculiarly open to exploitation. Into this zone crowded a varied assemblage of our early relatives.

"In habitat," comments one scholar, "many of these early primates may be thought of as the rats of the Paleocene. With the later appearance of true rodents, the primate habitat was markedly restricted." The bone hunters, in other words, have succeeded in demonstrating that numerous primates reveal a remarkable development of rodent-like characteristics in the teeth and skull during this early period of mammalian evolution. The movement is progressive and distributed in several different groups. One form, although that of a true primate, shows similarities to the modern kangaroo rat, which is, of course, a rodent. There is little doubt that it was a burrower.

It is this evidence of a lost chapter in the history of our kind that I used to remember on the sunny slope above prairie-dog town, and why I am able to say in a somewhat figuratively fashion that we were driven out of it once ages ago. We are not, except very remotely as mammals, related to prairie dogs. Nevertheless, through several millions years of Paleocene time, the primate order, instead of being confined to trees, was experimenting to some extent with the same grassland burrowing life that the rodents later perfected. The success of these burrowers crowded the primates out of this environment and forced them back into the domain of the branches. As a result, many primates, by that time highly specialized for a ground life, became extinct.

In the restricted world of the trees, a "refuge area," as the zoologist would say, the others lingered on in diminished numbers. Our ancient relatives, it appears, were beaten in their attempt to expand upon the ground; they were dying out in the temperate zone, and their significance as a widespread and diversified group was fading. The shabby pseudo-rat I had seen ruffling his coat to dry after the night damps of the reptile age, had ascended again into the green twilight of the rain forest. The chatterers with the ever-growing teeth were his masters. The sunlight and grass belonged to them.

It is conceivable that except for the invasion of the rodents, the primate line might even have abandoned the trees. We might be there in the grass, you and I, barking in the high-plains sunlight. It is true we came back in fifty million years with the cunning hands and the eyes that the tree world gave us, but was it victory? Once more in memory I saw the high blue evening fall sleepily upon that village, and once more swung the car to leave, lifting, as I always did, a figurative lantern to some ambiguous crossroads sign within my brain. The pointing arms were nameless and nameless were the distances to which they pointed. One took one's choice.

I ceased my daydreaming then, squeezed myself out of the crevice, shook out my pipe, and started chipping once more, the taps sounding along the inward-leaning walls of the Slit like the echo of many footsteps ascending and descending. I had come a long way down since morning; I had projected myself across a dimension I was not fitted to traverse in the flesh. In the end I collected my tools and climbed painfully up through the colossal debris of ages. When I put my hands on the surface of the crack I looked all about carefully in a sudden anxiety that it might not be a grazing horse that I would see.

He had not visibly changed, however, and I mounted in some slight trepidation and rode off, having a memory for a camp—if I had gotten a foot in the right era—which should lie somewhere over to the west. I did not, however, escape totally that brief imprisonment.

Perhaps the Slit, with its exposed bones and its far-off vanishing sky, has come to stand symbolically in my mind for a dimension denied to man, the dimension of time. Like the wisteria on the garden wall he is rooted in his particular century. Out of it—forward or backward—he cannot run. As he stands on his circumscribed pinpoint of time, his sight for the past is growing longer, and even the shadowy outlines of the galactic future are growing clearer, though his own fate he cannot yet see. Along the dimension of time, man, like the rooted vine in space, may never pass in person. Considering the innumerable devices by which the mindless root has evaded the limitations of its own stability, however, it may well be that man himself is slowly achieving powers over a new dimension—a dimension capable of presenting him with a wisdom he has barely begun to discern.

Through how many dimensions and how many media will life have to pass? Down how many roads among the stars must man propel himself in search of the final secret? The journey is difficult, immense, at times impossible, yet that will not deter some of us from attempting it. We cannot know all that has happened in the past, or the reason for all of these events, any more than we can with surety discern what lies ahead. We have joined the caravan, you might say, at a certain point; we will travel as far as we can, but we cannot in one lifetime see all that we would like to see or learn all that we hunger to know. ▪

Antoine de Saint-Exupéry

FROM *Wind, Sand and Stars* (1939)
   translated from the French by Lewis Galantière

But by the grace of the airplane I have known a more extraordinary experience than this, and have been made to ponder with even more bewilderment the fact that this earth that is our home is yet in truth a wandering star.

A minor accident has forced me down in the Rio de Oro region, in Spanish Africa. Landing on one of those table-lands of the Sahara which fall away steeply at the sides, I found myself on the flat top of the frustrum of a cone, an isolated vestige of a plateau that had crumbled round the edges. In this part of the Sahara such truncated cones are visible from the air every hundred miles or so, their smooth surfaces always at about the same altitude above the desert and their geologic substance always identical. The surface sand is composed of minute and distinct shells; but progressively as you dig along a vertical section, the shells become more fragmentary, tend to cohere, and at the base of the cone form a pure calcareous deposit.

Without question, I was the first human being ever to wander over this . . . this iceberg; its sides were remarkably steep, no Arab could have climbed them, and no European had as yet ventured into this wild region.

I was thrilled by the virginity of a soil which no step of man or beast had sullied. I lingered there, startled by this silence that never had been broken. The first star began to shine, and I said to myself that this pure surface had lain here thousands of years in sight only of the stars.

But suddenly my musings on this white sheet and these shining stars were endowed with a singular significance. I had kicked against a hard, black stone, the size of a man's fist, a sort of molded rock of lava incredibly present on the surface of a bed of shells a thousand feet deep. A sheet spread beneath an apple-tree can receive only apples; a sheet spread beneath the stars can receive only star-dust. Never had a stone fallen from the skies made known its origin so unmistakably.

And very naturally, raising my eyes, I said to myself that from the height of this celestial apple-tree there must have dropped other fruits, and that I should find them exactly where they fell, since never from the beginning of time had anything been present to displace them.

Excited by my adventure, I picked up one and then a second and then a third of these stones, finding them at about the rate of one stone to the acre. And here is where my adventure became magical, for in a striking foreshortening of time that embraced thousands of years, I had become the witness of this miserly rain from the stars. The marvel of marvels was that there on the rounded back of the planet, between this magnetic sheet and those stars, a human consciousness was present in which as in a mirror that rain could be reflected. ▪

Rita Dove

FROM *Museum* (1983)

THE FISH IN THE STONE

The fish in the stone
would like to fall
back into the sea.

He is weary
of analysis, the small
predictable truths.
He is weary of waiting
in the open,
his profile stamped
by white light.

In the ocean the silence
moves and moves

and so much is unnecessary!
Patient, he drifts
until the moment comes
to cast his
skeletal blossom.

The fish in the stone
knows to fail is
to do the living
a favor.

He knows why the ant
engineers a gangster's
funeral, garish
and perfectly amber.
He knows why the scientist
in secret delight
strokes the fern's
voluptuous Braille. ∎

James Hutton

FROM *Theory of the Earth* (1795)

If, in pursuing this object, we employ our skill in research, not in forming vain conjectures; and if *data* are to be found, on which Science may form just conclusions, we should not long remain in ignorance with respect to the natural history of this earth, a subject on which hitherto opinion only, and not evidence, has decided: For in no subject, perhaps, is there naturally less defect of evidence, although philosophers, led by prejudice, or mis-guided by false theory, may have neglected to employ that light by which they should have seen the system of this world.

But to proceed in pursuing a little farther our general or preparatory ideas. A solid body of land could not have answered the purpose of a hab-itable world; for, a soil is necessary to the growth of plants; and a soil is nothing but the materials collected from the destruction of the solid land. Therefore, the surface of this land, inhabited by man, and covered with plants and animals, is made by nature to decay, in dissolving from that hard and compact state in which it is found below the soil; and this soil is necessarily washed away, by the continual circulation of the water, run-ning from the summits of the mountains towards the general receptacle of that fluid.

The heights of our land are thus levelled with the shores; our fertile plains are formed from the ruins of the mountains; and those travelling materials are still pursued by the moving water, and propelled along the inclined surface of the earth. These moveable materials, delivered into the sea, cannot, for a long continuance, rest upon the shore; for, by the agita-tion of the winds, the tides and currents, every moveable thing is carried farther and farther along the shelving bottom of the sea, towards the unfathomable regions of the ocean.

If the vegetable soil is thus constantly removed from the surface of the land, and if its place is thus to be supplied from the dissolution of the solid earth, as here represented, we may perceive an end to this beautiful machine; an end, arising from no error in its constitution as a world, but from that destructibility of its land which is so necessary in the system of the globe, in the economy of life and vegetation.

The immense time necessarily required for this total destruction of the land, must not be opposed to that view of future events, which is indicated by the surest facts, and most approved principles. Time, which measures every thing in our idea, and is often deficient to our schemes, is to nature endless and as nothing; it cannot limit that by which alone it had existence; and, as the natural course of time, which to us seems infinite, cannot be bounded by any operation that may have an end, the progress of things upon this globe, that is, the course of nature, cannot be limited by time, which must proceed in a continual succession. We are, therefore, to consider as inevitable the destruction of our land, so far as effected by those operations which are necessary in the purpose of the globe, considered as a habitable world; and, so far as we have not examined any other part of the economy of nature, in which other operations and a different intention might appear.

We have now considered the globe of this earth as a machine, constructed upon chemical as well as mechanical principles, by which its different parts are all adapted, in form, in quality, and in quantity, to a certain end; an end attained with certainty or success; and an end from which we may perceive wisdom, in contemplating the means employed.

But is this world to be considered thus merely as a machine, to last no longer than its parts retain their present position, their proper forms and qualities? Or may it not be also considered as an organized body? Such as has a constitution in which the necessary decay of the machine is naturally repaired, in the exertion of those productive powers by which it had been formed.

This is the view in which we are now to examine the globe; to see if there be, in the constitution of this world, a reproductive operation, by which a ruined constitution may be again repaired, and a duration or stability thus procured to the machine, considered as a world sustaining plants and animals.

[. . .]

To sum up the argument, we are certain, that all the coasts of the present continents are wasted by the sea, and constantly wearing away, upon the whole; but this operation is so extremely slow, that we cannot find a measure of the quantity in order to form an estimate: Therefore, the present continents of the earth, which we consider as in a state of perfection, would, in the natural operations of the globe, require a time indefinite for their destruction.

But, in order to produce the present continents, the destruction of a former vegetable world was necessary; consequently, the production of our present continents must have required a time which is indefinite. In like manner, if the former continents were of the same nature as the present, it must have required another space of time, which also is indefinite, before they had come to their perfection as a vegetable world.

We have been representing the system of this earth as proceeding with a certain regularity, which is not perhaps in nature, but which is necessary for our clear conception of the system of nature. The system of nature is certainly in rule, although we may not know every circumstance of its regulation. We are under a necessity, therefore, of making regular suppositions, in order to come at certain conclusions which may be compared with the present state of things.

[. . .]

In thus accomplishing a certain end, we are not to limit nature with the uniformity of an equable progression, although it be necessary in our computations to proceed upon equalities. Thus also, in the use of means, we are not to prescribe to nature those alone which we think suitable for the purpose, in our narrow view. It is our business to learn of nature (that is by observation) the ways and means, which in her wisdom are adopted; and we are to imagine these only in order to find means for further information, and to increase our knowledge from the examination of things which actually have been. It is in this manner, that intention may be found in nature; but this intention is not to be supposed, or vainly imagined, from what we may conceive to be.

[. . .]

We have now got to the end of our reasoning; we have no data further to conclude immediately from that which actually is: But we have got enough; we have the satisfaction to find, that in nature there is wisdom, system, and consistency. For having, in the natural history of this earth, seen a succession of worlds, we may from this conclude that there is a system in nature; in like manner as, from seeing revolutions of the planets, it is concluded, that there is a system by which they are intended to continue those revolutions. But if the succession of worlds is established in the system of nature, it is in vain to look for any thing higher in the origin of the earth. The result, therefore, of this physical inquiry is, that we find no vestige of a beginning,—no prospect of an end. ■

Paul Shepard

FROM *Man in the Landscape* (1967)

Perhaps there is no better example of the evocative power of natural land-scapes than the response of westering pioneers to novel erosional remnants and angular cliffs. To many of the thousands who followed the Oregon Trail before 1850, the escarpments and sedimentary bluffs along the Platte River in western Nebraska were the structures of a ghostly architecture. The Reverend Samuel Parker wrote in his diary in 1835:

> Encamped today near what I shall call the old castle, which is a great natural curiosity . . . [it has] the appearance of an old enormous build-ing, somewhat dilapidated; but still you see the standing walls, the roof, the turrets, embrasures, the dome, and almost the very windows; and the guard houses, large, and standing some rods in front of the main building. You unconsciously look around for the enclosures but they are all swept away by the lapse of time—for the inhabitants, but they have disappeared; all is silent and solitary . . .

These speculations were more than the whimsy of a saddlesore preacher. The journals of mountain men, farmers, speculators, and soldiers are replete with similar comparisons. Their wonder is directed towards "rocks" called Steamboat, Table, Castle, Smokestack, Roundhouse, Court-house, Jail, and Chimney. Why should these particular rocks have looked more like buildings than any back East? And why should they have made such an indelible impression on the traveler? These are problems in human ecology, of the formation of attitudes toward the landscape; of the fusion of an experience in nature with historical ideas of process and natural change; and they reveal a projection into new situations of values evolved in an old, familiar, and different environment.

Proceeding northwest along the terraced banks of the Platte at a longi-tude just beyond 103 degrees, a few minutes south of the 42nd parallel, the traveler was about six weeks out of Independence, Missouri. After cross-ing the upland between the Platte forks, he had followed for almost two weeks the "shores" of what Washington Irving described as the "most beautiful and least useful" river in America. Because the Platte was a kind of linear oasis the itinerant was scarcely aware of the progressive alteration in vegetation and land forms. The travelers hailed from several states, particularly Eastern and Midwestern. They were heterogeneous groups, numbering sometimes in the thousands. They were alike insofar as they shared the geological provincialism of men reared in the sub-humid forest landscapes of America and Europe. They shared also the historical background and values of Protestant Yankees and Hoosiers of the 1830's and 1840's.

At Scott's Bluff the itinerant had climbed more than three thousand feet above Independence. He had traversed the northern high plains from the oak-hickory forest to the margin of semiarid highlands, from regions of more than thirty inches of rainfall annually to under fifteen. Leaving the savannahs of the western boundaries of the forests, he had crossed the tall-grass prairies and the shorter mixed grasses, to the place where the upland vegetational cover ceased to be continuous—a significant point in the vegetational influence on the geomorphic processes of weathering and mass wasting. The traveler had also entered a region of greater daily temperature range, more numerous cyclonic storms, and less relative humidity, with their varied and subtle effects on human perception and response.

Plodding up the valley of the Platte, with its arm of forest, meadow, and savannah, the traveler penetrated unaware new biotic and geomorphic surroundings. Shortly after he had entered what is now Scott's Bluff County, Jailhouse Rock and Courthouse Rock appeared on the left horizon some fifteen miles away. As the column passed slowly to the right of these structures, more came into view, finally an escarpment parallel to the trail about five miles from the river. This mountain, Wildcat Ridge on present maps, is more than thirty miles long and sends three spurs north almost to the river's edge, the westernmost being Scott's Bluff.

The first fifteen miles or so produced a galvanizing impact on the observer. There had been intimations of things to come, such as buffalo trails that looked to one pioneer "like the once oft-trodden streets of some

deserted city." The valley with its pleasant greenery had itself been sug-
gestive; Rufus B. Sage observed that "everything has more of the appearance
of civilization than anything that I have seen for many days, the trees, the
shrubs and bushes, grapevines, the grass—resembling blue grass—the
singing of the birds in the trees, the sound of the ax cutting wood for breakfast . . ." Then, as the westbound party drew abreast of the bluffs, a wave of
astonishment swept through it. John Bidwell wrote in 1841, "the scenery
of the surrounding country became beautifully grand and picturesque—
they were worn in such a manner by the storm of unnumbered seasons, that
they really counterfeited the lofty spires, towering edifices, spacious domes,
and in fine all the beautiful mansions of cities."

Numerous observers discovered lighthouses, brick kilns, the Capitol of
Washington, Beacon Hill, shot towers, churches, spires, cupolas, streets,
workshops, stores, warehouses, parks, squares, pyramids, castles, forts, pillars, domes, minarets, temples, Gothic castles, "modern" fortifications,
French cathedrals, Rhineland castles, towers, tunnels, hallways, mau-
soleums, a Temple of Belus, and hanging gardens which were "in a tolerable state of preservation, and showing in many places hardy shrubs that,
having sent down their long roots into partial openings of the supporting
arches, still smiled in beautiful green, amid general desolation," according
to J. Quinn Thornton. Taken at a glance the rocks "had the appearance of
Cities, Temples, Castles, Towers, Palaces, and every variety of great and
magnificent structures . . . splendid edifices, like beautiful white marble,
fashioned in the style of every age and country," reported Overton Johnston
and William Winter. Where more palpably than in America could such a
jumble of architecture actually look like a city?

> Here were the minarets of a castle; there the loopholes of bastions of
> a fort; again the frescoes of a huge temple; there the doors, windows,
> chimneys, and the columns of immense buildings appeared in view,
> with all the solemn grandeur of an ancient, yet deserted city, while
> at other points Chinese temples, dilapidated by time, broken chim-
> neys, rocks in miniature made it appear as if by some supernatural
> cause we had been dropped in the suburbs of a mighty city—for miles

around the basin this view extended, and we looked across the barren plain at the display of Almighty power, with wonder and astonishment. [A. Delano]

But the cities were not often American. What cities came to mind?

The mind was filled with strange images and impressions. The silence of death reigned over a once populous city, which had been a nursery of the arts and sciences, and the seat of a grand inland commerce. It was a Tadmore of the desert in ruins. [J. Q. Thornton]

What people had lived there?

No effort of the imagination is required to suppose ourselves encamped in the vicinity of the ruins of some vast city erected by a race of giants, contemporaries of the Megatherii and Icthyosaurii. [Edwin Bryant]

Noble castles with turrets, embrasures, and loopholes, with drawbridge in front and the moat surrounding it; behind, the humble cottages of the subservient peasantry and all the varied concomitants of such a scene, are so strikingly evident to the view, that it required but little stretch of fancy to imagine that a race of antediluvian giants may have here swayed their iron sceptre, and left behind the crumbling palace and the tower, to all of their departed glory. [John K. Townsend]

What had happened to them? There was a room, suggested J.Q. Thornton, where "that monarch might have sat upon his throne, surrounded by obsequious courtiers and servile slaves, while the lifeblood of men better than himself was being shed to make him a holiday." Perhaps because of this degeneracy the city had been overwhelmed. Another suggested that it had been occupied by "a people who had perhaps gone down into the vortex of revolutions . . . leaving no trace of their existence, save those remains of architectural grandeur and magnificence." From the position of the ruins some travelers reconstructed the probable course of the catastrophe, a series of pitched battles, slaughter, pillage, fire, and the "bodies in promiscuous piles about the gates."

The illusion was so difficult to resist that a present reader of these journals discriminates with difficulty between a speculative visual play on forms and their animation by ghosts from the European and Biblical past. The mirage "would deceive the most practiced eye were it not known that it is situated in a wilderness hundreds of miles from any habitation." There was a continual protestation of bemusement and flashes of embarrassed self-consciousness. The Reverend Sam Parker declared that "one can hardly believe that they are not the work of art. Although you correct your imagination, and call to remembrance, that you are beholding the work of nature, yet before you are aware, the illusion takes you again, and again your curiosity is excited to know who built this fabric, and what has become of the bygone generations." ▪

Mark Twain

FROM *Letters from the Earth*
  edited by Bernard DeVoto

"Was the World Made for Man?"

I seem to be the only scientist and theologian still remaining to be heard from on this important matter of whether the world was made for man or not. I feel that it is time for me to speak.

I stand almost with the others. They believe the world was made for man, I believe it likely that it was made for man; they think there is proof, astronomical mainly, that it was made for man, I think there is evidence only, not proof, that it was made for him. It is too early, yet, to arrange the verdict, the returns are not all in. When they are all in, I think they will show that the world was made for man; but we must not hurry, we must patiently wait till they are all in.

Now as far as we have got, astronomy is on our side. Mr. Wallace has clearly shown this. He has clearly shown two things: that the world was made for man, and that the universe was made for the world—to stiddy it, you know. The astronomy part is settled, and cannot be challenged.

We come now to the geological part. This is the one where the evidence is not all in, yet. It is coming in, hourly, daily, coming in all the time, but naturally it comes with geological carefulness and deliberation, and we must not be impatient, we must not get excited, we must be calm, and wait. To lose our tranquillity will not hurry geology; nothing hurries geology.

It takes a long time to prepare a world for man, such a thing is not done in a day. Some of the great scientists, carefully ciphering the evidences furnished by geology, have arrived at the conviction that our world is prodigiously old, and they may be right, but Lord Kelvin is not of their opinion. He takes a cautious, conservative view, in order to be on the safe side, and feels sure it is not so old as they think. As Lord Kelvin is the highest authority in science now living, I think we must yield to him and accept his view. He does not concede that the world is more than a hundred million years old. He believes it is that old, but not older. Lyell believed that our race was introduced into the world 31,000 years ago, Herbert Spencer makes it 32,000. Lord Kelvin agrees with Spencer.

Very well. According to these figures it took 99,968,000 years to prepare the world for man, impatient as the Creator doubtless was to see him and admire him. But a large enterprise like this has to be conducted warily, painstakingly, logically. It was foreseen that man would have to have the oyster. Therefore the first preparation was made for the oyster. Very well, you cannot make an oyster out of whole cloth, you must make the oyster's ancestor first. This is not done in a day. You must make a vast variety of invertebrates, to start with—belemnites, trilobites, Jebusites, Amalekites, and that sort of fry, and put them to soak in a primary sea, and wait and see what will happen. Some will be a disappointment—the belemnites, the Ammonites and such; they will be failures, they will die out and become extinct, in the course of the nineteen million years covered by the experiment, but all is not lost, for the Amalekites will fetch the homestake; they will develop gradually into encrinites, and stalactites, and blatherskites, and one thing and another as the mighty ages creep on and the Archaean and the Cambrian Periods pile their lofty crags in the primordial seas, and at last the first grand stage in the preparation of the world for man stands completed, the oyster is done. An oyster has hardly any more reasoning power than a scientist has; and so it is reasonably certain that this one jumped to the conclusion that the nineteen million years was a preparation for *him*; but that would be just like an oyster, which is the most conceited animal there is, except man. And anyway, this one could not know, at that early date, that he was only an incident in a scheme, and that there was some more to the scheme, yet.

The oyster being achieved, the next thing to be arranged for in the preparation of the world for man was fish. Fish, and coal—to fry it with. So the Old Silurian seas were opened up to breed the fish in, and at the same time the great work of building Old Red Sandstone mountains eighty thousand feet high to cold-storage their fossils in was begun. This latter was quite indispensable, for there would be no end of failures again, no end of extinctions—millions of them—and it would be cheaper and less trouble to can them in the rocks than keep tally of them in a book. One does not build the coal beds and eighty thousand feet of perpendicular Old Red Sandstone in a brief time—no, it took twenty million years. In the first place, a coal bed is a slow and troublesome and tiresome thing to construct. You have to grow prodigious forests of tree-ferns and reeds and calamites and such things in a marshy region; then you have to sink them under out of sight and let them rot; then you have to turn the streams on them, so as to bury them under several feet of sediment, and the sediment must have time to harden and turn to rock; next you must grow another forest on top, then sink it and put on another layer of sediment and harden it; then more forest and more rock, layer upon layer, three miles deep—ah, indeed it is a sickening slow job to build a coal-measure and do it right!

So the millions of years drag on; and meantime the fish culture is lazying along and frazzling out in a way to make a person tired. You have developed ten thousand kinds of fishes from the oyster; and come to look, you have raised nothing but fossils, nothing but extinctions. There is nothing left alive and progressive but a ganoid or two and perhaps half a dozen asteroids. Even the cat wouldn't eat such.

Still, it is no great matter; there is plenty of time, yet, and they will develop into something tasty before man is ready for them. Even a ganoid can be depended on for that, when he is not going to be called on for sixty million years.

The Paleozoic time limit having now been reached, it was necessary to begin the next stage in the preparation of the world for man, by opening up the Mesozoic Age and instituting some reptiles. For man would need reptiles. Not to eat, but to develop himself from. This being the most important detail of the scheme, a spacious liberality of time was set apart for it— thirty million years. What wonders followed! From the remaining ganoids and asteroids and alkaloids were developed by slow and steady and painstaking culture those stupendous saurians that used to prowl about the steamy world in those remote ages, with their snaky heads reared forty feet in the air and sixty feet of body and tail racing and thrashing after. All gone, now, alas—all extinct, except the little handful of Arkansawrians left stranded and lonely with us here upon this far-flung verge and fringe of time.

Yes, it took thirty million years and twenty million reptiles to get one that would stick long enough to develop into something else and let the scheme proceed to the next step.

Then the pterodactyl burst upon the world in all his impressive solemnity and grandeur, and all Nature recognized that the Cenozoic threshold was crossed and a new Period open for business, a new stage begun in the preparation of the globe for man. It may be that the pterodactyl thought the thirty million years had been intended as a preparation for himself, for there was nothing too foolish for a pterodactyl to imagine, but he was in error, the preparation was for man. Without doubt the pterodactyl attracted great attention, for even the least observant could see that there was the making of a bird in him. And so it turned out. Also the makings of a mammal, in time. One thing we have to say to his credit, that in the matter of picturesqueness he was the triumph of his Period; he wore wings and had teeth, and was a starchy and wonderful mixture altogether, a kind of long-distance premonitory symptom of Kipling's marine:

> 'E isn't one o' the reg'lar Line, nor 'e isn't one of the crew,
> 'E's a kind of a giddy harumfrodite—soldier an' sailor too!

From this time onward for nearly another thirty million years the preparation moved briskly. From the pterodactyl was developed the bird; from the bird the kangaroo, from the kangaroo the other marsupials; from these the mastodon, the megatherium, the giant sloth, the Irish elk, and all that crowd that you make useful and instructive fossils out of—then came the first great Ice Sheet, and they all retreated before it and crossed over the bridge at Bering Strait and wandered around over Europe and Asia and died. All except a few, to carry on the preparation with. Six Glacial Periods with two millions years between Periods chased these poor orphans up and down and about the earth, from weather to weather—from tropic swelter at the poles to Arctic frost at the equator and back again and to and fro, they never knowing what kind of weather was going to turn up next; and if ever they settled down anywhere the whole continent suddenly sank under them without the least notice and they had to trade places with the fishes and scramble off to where the seas had been, and scarcely a dry rag on them; and when there was nothing else doing a volcano would let go and fire them out from wherever they had located. They led this unsettled and irritating life for twenty-five million years, half the time afloat, half the time aground, and always wondering what it was all for, they never suspecting, of course, that it was a preparation for man and had to be done just so or it wouldn't be any proper and harmonious place for him when he arrived.

And at last came the monkey, and anybody could see that man wasn't far off, now. And in truth that was so. The monkey went on developing for close upon five million years, and then turned into a man—to all appearances.

Such is the history of it. Man has been here 32,000 years. That it took a hundred million years to prepare the world for him is proof that that is what it was done for. I suppose it is. I dunno. If the Eiffel Tower were now representing the world's age, the skin of paint on the pinnacle-knob at its summit would represent man's share of that age; and anybody would perceive that that skin was what the tower was built for. I reckon they would, I dunno. ▪

3 Faults, Earthquakes, and Tsunamis

By following cracks you can trace the subtle power
of the fault as it angles under the town, offsetting sidewalks and curbstones and gutters, an effect most alarming in the house of a chiropractor which you pass soon
after entering Hollister from the west. One half of a low
concrete retaining wall holding back the chiropractor's
lawn has been carried north and west about eight inches.
The concrete walkway is buckling. Both porch pillars
lean precariously toward the coast. In back, the wall
of his garage is bent into a curve like a stack of whale's
ribs. The fact that half his doomed house rides on the
American plate and the other half rides the Pacific has
not discouraged this chiropractor from maintaining a
little order in his life. He hangs his sign out front, he
keeps his lawn well mowed and the old house brightly,
spotlessly painted.

—James D. Houston, *Continental Drift*

SOME OF OUR OLDEST STORIES AND WRITINGS INCLUDE ATTEMPTS TO describe or explain abrupt, violent ground shaking. Proposed explanations of earthquakes have included divine warnings, internal explosions, and, in the current field of seismology, sudden releases of stored energy along faults in the Earth.

Seismologists note that the pattern of earthquake occurrence illustrates the dynamic nature of our planet by marking the boundaries of tectonic plates, much like the seams of an earthly baseball. Along these boundaries the Earth's tectonic plates split and pull apart, grind alongside, or converge with each other. According to current geologic theory, these plate motions are the surface result of slow continuous movement—about the rate of growth of a fingernail—of hot solid rock within the Earth's interior. But closer to Earth's surface, within its top few tens to hundreds of kilometers, the motions are jerky; rocks rupture, abruptly shift past each other, and cause earthquakes.

Some of the stored-up energy released in an earthquake causes vibrations or seismic waves that travel through the planet and/or along its surface. These waves are of several different types, each with particular styles of motion and rates of travel. The differences in wave type affecting various areas account for the ranges of experiences and descriptions of earthquakes. In addition, the nature of the faulting determines the type of earthquake.

The "Ring of Fire" surrounding the Pacific Ocean is one the most important and active seismic belts in the world. Most of the earthquakes there occur along convergent boundaries, where plates converge and one plate dives beneath the other. As the plates descend back down into Earth's interior, some of the rocks melt and cause active volcanoes to form. Such mountains born of fire are considered in part 4. Another major region of earthquake occurrence extends from the Straits of Gibraltar through southern Europe, western Asia, the Himalaya, China, and southeast Asia where it joins the "Ring of Fire." This zone of earthquakes marks the plate convergence and collision of Africa, the Arabian Peninsula, India, and Australia with Eurasia.

Earthquakes are often terrifying to experience. The noise and the shaking of "solid" ground beneath one's feet can damage or destroy any innate sense of "terra firma" as an immovable foundation for our existence. Yet earthquakes in and of themselves seldom cause injury or death. Rather, damage and loss of life from earthquakes often directly result from building collapse, fire caused by broken gas or electrical lines, tsunamis (or seismic sea waves) in coastal areas, landslides, and dam failure.

This section draws on writings from the twelfth century to the present day, from Japan and the west coast of North America. In eighteenth-century Europe, the balance of opinion on the causes of earthquakes swayed between the prevailing clerical view of their occurrence as the ultimate expression of Divine wrath—of warnings to and punishments of unrepentant sinners—and the budding scientific view of earthquakes as a manifestation of still poorly understood interior workings of the planet. In addition to words on fault theory and geography, and the cause of earthquakes, the writings in this chapter present experiences of terror and destruction, responses to disaster, and personal reflections on coming to terms with what is ultimately unknowable and uncontrollable.

All of this section's excerpts concern earthquakes and faults in the Pacific "Ring of Fire." We open with a vivid account by Kamo no Chomei (1153– 1216) of a large earthquake that devastated part of Japan in 1185. From "earthquake country" in California we include several accounts of tremors on the San Andreas Fault system, which constitutes about half the length of the boundary between the North American and Pacific plates. In an excerpt from *Roughing It,* Mark Twain (Samuel Clemens, 1835–1910) gives a humorous account of the 1868 earthquake, which occurred on the Hayward Fault, a branch of the San Andreas Fault system, near Berkeley, California. The 1868 earthquake was considered the "great earthquake" until 1906.

We selected two eyewitness accounts of the great California earthquake of 1906, which is estimated to have released about twenty-five times as much energy as the 1868 event. This earthquake occurred at 5:12 A.M. on April 8, 1906, along the San Andreas Fault in the San Francisco Bay Area. With an estimated Richter magnitude of 8.3, this tremor caused an estimated seven hundred deaths and $4 million in property damage—with about 70 percent of the damage due to fire. Grove Karl Gilbert (1843–1918), a foremost geologist of his time and one of the greatest American geologists ever, was one of the first to realize that earthquakes occurred because of slip along faults; prior to that time, faults were thought by geologists to be secondary effects of earthquakes. The intense personal detail of Mary Austin's account of the events that day in the city of 400,000 people gives a sense of the unsettling fear and tragedy felt by all.

These pieces are followed by a selection from "Assembling California" in *Annals of the Former World,* in which John McPhee describes the distinct landscape patterns left by movement along the San Andreas Fault in California.

The section ends with a report by Eliza Ruhamah Scidmore (for *National Geographic Magazine*) on the tsunami that struck the coast of Hondo, the main island of Japan, in 1896.

Kamo no Chomei

FROM *An Account of My Hut* (1212)
    translated from the Japanese by Donald Keene (1955)

The Earthquake

Then there was the great earthquake of 1185, of an intensity not known before. Mountains crumbled and rivers were buried, the sea tilted over and immersed the land. The earth split and water gushed up; boulders were sundered and rolled into the valleys. Boats that rowed along the shores were swept out to sea. Horses walking along the roads lost their footing. It is needless to speak of the damage throughout the capital—not a single mansion, pagoda, or shrine was left whole. As some collapsed and others tumbled over, dust and ashes rose like voluminous smoke. The rumble of the earth shaking and the houses crashing was exactly like that of thunder. Those who were in their houses, fearing that they would presently be crushed to death, ran outside, only to meet with a new cracking of the earth. They could not soar into the sky, not having wings. They could not climb into the clouds, not being dragons. Of all the frightening things of the world, none is so frightful as an earthquake.

Among those who perished was the only child of a samurai family, a boy of five or six, who had made a little house under the overhanging part of a wall and was playing there innocently when the wall suddenly collapsed, burying him under it. His body was crushed flat, with only his two eyes protruding. His parents took him in their arms and wailed uncontrollably, so great was the sorrow they experienced. I realized that grief over a child can make even the bravest warrior forget shame—a pitiable but understandable fact.

The intense quaking stopped after a time, but the after-tremors continued for some while. Not a day passed without twenty or thirty tremors of a severity which would ordinarily have frightened people. After a week or two their frequency diminished, and there would be four or five, then two or three a day; then a day might be skipped, or there would be only one tremor in two or three days. After-tremors continued for three months.

Of the four great elements, water, fire, and wind are continually causing disasters, but the earth does not normally afflict man. Long ago, during the great earthquake of the year 855, the head of the Buddha of the Tōdaiji fell off, a terrible misfortune, indeed, but not the equal of the present disaster. At the time everyone spoke of the vanity and meaninglessness of the world, and it seemed that the impurities in men's hearts had somewhat lessened, but with the passage of the months and the days and the coming of the new year people no longer even spoke in that vein. ▪

Mark Twain

FROM *Roughing It* (1872)

A month afterward I enjoyed my first earthquake. It was one which was
long called the "great" earthquake, and is doubtless so distinguished till
this day. It was just after noon, on a bright October day. I was coming down
Third street. The only objects in motion anywhere in sight in that thickly
built and populous quarter, were a man in a buggy behind me, and a street
car wending slowly up the cross street. Otherwise, all was solitude and a
Sabbath stillness. As I turned the corner, around a frame house, there was
a great rattle and jar, and it occurred to me that here was an item!—no
doubt a fight in that house. Before I could turn and seek the door, there
came a really terrific shock; the ground seemed to roll under me in waves,
interrupted by a violent joggling up and down, and there was a heavy
grinding noise as of brick houses rubbing together. I fell up against the
frame house and hurt my elbow. I knew what it was, now, and from mere
reportorial instinct, nothing else, took out my watch and noted the time
of day; at that moment a third and still severer shock came, and as I reeled
about on the pavement trying to keep my footing, I saw a sight! The entire
front of a tall four-story brick building in Third Street sprung outward like
a door and fell sprawling across the street, raising a dust like a great volume of smoke! And here came the buggy—overboard went the man, and
in less time than I can tell it the vehicle was distributed in small fragments
along three hundred yards of street. One could have fancied that somebody
had fired a charge of chair-rounds and rags down the thoroughfare. The
street car had stopped, the horses were rearing and plunging, the passengers were pouring out at both ends, and one fat man had crashed half way
through a glass window on one side of the car, got wedged fast and was
squirming and screaming like an impaled madman. Every door, of every
house, as far as the eye could reach, was vomiting a stream of human
beings; and almost before one could execute a wink and begin another,
there was a massed multitude of people stretching in endless procession

down every street my position commanded. Never was solemn solitude turned into teeming life quicker.

Of the wonders wrought by "the great earthquake," these were all that came under my eye; but the tricks it did, elsewhere, and far and wide over the town, made toothsome gossip for nine days. The destruction of property was trifling—the injury to it was widespread and somewhat serious.

The "curiosities" of the earthquake were simply endless. Gentlemen and ladies who were sick, or were taking a siesta, or had dissipated till a late hour and were making up lost sleep, thronged into the public streets in all sorts of queer apparel, and some without any at all. One woman who had been washing a naked child, ran down the street holding it by the ankles as if it were a dressed turkey. Prominent citizens who were supposed to keep the Sabbath strictly, rushed out of saloons in their shirt-sleeves, with billiard cues in their hands. Dozens of men with necks swathed in napkins, rushed from barber-shops, lathered to the eyes or with one cheek clean shaved and the other still bearing a hairy stubble. Horses broke from stables, and a frightened dog rushed up a short attic ladder and out on to a roof, and when his scare was over had not the nerve to go down again the same way he had gone up. A prominent editor flew down stairs, in the principal hotel, with nothing on but one brief undergarment—met a chambermaid, and exclaimed:

"Oh, what *shall* I do! Where shall I go!"

She responded with naive serenity:

"If you have no choice, you might try a clothing-store!"

A certain foreign consul's lady was the acknowledged leader of fashion, and every time she appeared in anything new or extraordinary, the ladies in the vicinity made a raid on their husbands' purses and arrayed themselves similarly. One man who had suffered considerably and growled accordingly, was standing at the window when the shocks came, and the next instant the consul's wife, just out of the bath, fled by with no other apology for clothing than—a bath-towel! The sufferer rose superior to the terrors of the earthquake, and said to his wife:

"Now *that* is something *like!* Get out your towel my dear!"

The plastering that fell from ceilings in San Francisco that day, would have covered several acres of ground. For some days afterward, groups of eyeing and pointing men stood about many a building, looking at long zig-zag cracks that extended from the eaves to the ground. Four feet of the tops of three chimneys on one house were broken square off and turned around in such a way as to completely stop the draft. A crack a hundred feet long gaped open six inches wide in the middle of one street and then shut together again with such force, as to ridge up the meeting earth like a slender grave. A lady sitting in her rocking and quaking parlor, saw the wall part at the ceiling, open and shut twice, like a mouth, and then—drop the end of a brick on the floor like a tooth. [. . .]

The first shock brought down two or three huge organ-pipes in one of the churches. The minister, with uplifted hands, was just closing the services. He glanced up, hesitated, and said:

"However, we will omit the benediction!"—and the next instant there was a vacancy in the atmosphere where he had stood.

After the first shock, an Oakland minister said:

"Keep your seats! There is no better place to die than this"—

And added, after the third:

"But outside is good enough!" He then skipped out at the back door. ▪

Grove Karl Gilbert

FROM "The Investigation of the California Earthquake of 1906" (1906–7)

It is the natural and legitimate ambition of a properly constituted geologist to see a glacier, witness an eruption and feel an earthquake. The glacier is always ready, awaiting his visit; the eruption has a course to run, and alacrity only is needed to catch its more important phases; but the earthquake, unheralded and brief, may elude him through his entire lifetime. It had been my fortune to experience only a single weak tremor, and I had, moreover, been tantalized by narrowly missing the great Inyo earthquake of 1872 and the Alaska earthquake of 1899. When, therefore, I was awakened in Berkeley on the eighteenth of April last by a tumult of motions and noises, it was with unalloyed pleasure that I became aware that a vigorous earthquake was in progress. The creaking of the building, which has a heavy frame of redwood, and the rattling of various articles of furniture so occupied my attention that I did not fully differentiate the noises peculiar to the earthquake itself. The motions I was able to analyze more successfully, perceiving that, while they had many directions, the dominant factor was a swaying in the north-south direction, which caused me to roll slightly as I lay with my head toward the east. Afterward I found a suspended electric lamp swinging in the north-south direction, and observed that water had been splashed southward from a pitcher. These notes of direction were of little value, however, except as showing control by the structure of the building, for in another part of the same building the east-west motion was dominant.

In my immediate vicinity the destructive effects were trivial, and I did not learn until two hours later that a great disaster had been wrought on the opposite side of the bay and that San Francisco was in flames. This information at once incited a tour of observation, and thus began, so far as I was personally concerned, the investigation of the earthquake. A similar beginning was doubtless made by every other geologist in the State, and the initial work of observation and record was individual and without concert. But organization soon followed, and by the end of the second day it is probable that twenty men were working in co-operation under the leadership of Professor J. C. Branner, of Stanford University, and Professor A. C. Lawson, of the State University at Berkeley. At that time and for several succeeding days the ordinary means of communication were so paralyzed or overburdened that no messages passed between these two centers of organization; but as the needs of the hour were patent to all, the work was not prejudiced by the lack of intercommunication.

[. . .]

Architects and engineers were not less prompt and energetic. To the men who plan and direct construction in the earthquake district of California it was important to know what materials and what structural forms best withstood the shock, and they immediately began the study of earthquake injuries and of instances of immunity from earthquake effects. In that part of San Francisco where the earthquake injury was most serious the shock was quickly followed by fire, which destroyed much of the evidence, but many important observations were made in the brief interval. The study of structural questions, like the study of natural phenomena, was at first individual only, but afterward was aided by organization. Committees were appointed by various professional societies, national and local, and were charged with the investigation of specific structural questions, and the results of their labors will find place not only in the transactions of the societies, but in revised building regulations and in important modifications of municipal plants for lighting and water supply. Various bureaus of the national government have also taken part in the structural studies, sending experts to San Francisco and other localities of exceptional earthquake violence.

The Japanese government promptly sent to California a committee of investigation headed by Dr. Omori, professor of seismology in the University of Tokyo, and composed otherwise of architects and engineers. The first conference of these visitors with the State Commission warranted the suggestion that we may find it as profitable to follow Japanese initiative in the matter of earthquake-resisting construction as in that of army hygiene.

[. . .]

The California earthquake was caused by a new slipping on the plane of an old fault which had been recognized for a long distance in California, and in one place had been named the San Andreas fault. Associated with this fault is a belt of peculiar topography, differing from the ordinary topographic expression of the country in that many of its features are directly due to dislocation, instead of being the product of erosion by rains and streams. One of its characteristics is the frequent occurrence of long lines of very straight cliffs. Another is the frequent occurrence of ponds or lakes in straight rows. The tendency of erosion is to break up such cliffs into series of spurs and valleys and to obliterate the lakes by cutting down their outlets or filling their basins with sediment. [. . .] This line and zone have been recognized by California geologists through a distance of several hundred miles. It was to this line that attention and expectation were especially directed, and it was on this line that the surface evidence of new faulting was actually found. The new movement was not coextensive with the line as previously traced, but affected only the northwestern portion; and, on the other hand, it extended farther to the northwest and north than the old line had previously been recognized. The map represents only the line along which the recent change occurred. From a point a few miles southwest of Hollister it runs northwestward in a series of valleys between low mountain ridges to the Mussel Rock, ten miles south of the Golden Gate. Thence northwestward and northward it follows the general coast line, alternately traversing land and water. The farthest point as yet definitely located is at Point Delgada, but the intensity of the shock at the towns of Petrolia and Ferndale probably indicates the close proximity of the fault and warrants the statement that its full length is not less than three hundred miles. South of Point Arena its course is direct, with only gentle flexure, but the data farther north seem to imply either branching or strong inflexion. Opposite San Francisco its position is several miles west of the coast line, and it nowhere touches a large town.

[. . .]

Wherever the shock was specially strong there were considerable injuries to trees; some were overturned, others broken near the ground, and yet others broken near their tops. A number of large redwood trees standing on the line of the rift were split from the ground upward, the basal portions being faulted along with the ground they stood on.

In the systematic survey of the earthquake area the relative intensity is being estimated by means of the records of various physical effects. In the immediate vicinity of the fault road-cracks and cracks in alluvium are large and numerous; many trees were broken or overturned; there were many landslides; half of the wooden buildings of any village or hamlet were shifted horizontally, often with serious injury; buildings and chimneys of brick or stone were thrown down; during the shock men, cows, and horses found it impossible to stand, and fell to the ground; and some persons were even thrown from their beds. In a general way all these evidences of violence diminish gradually with distance from the fault on either side. The rate of diminution, with exceptions to be mentioned presently, may be expressed by saying that at five miles from the fault only a few men and animals were shaken from their feet, only a few wooden houses were moved from their foundations, about half the brick chimneys remained sound and in condition for use, sound trees were not broken, and no cracks were opened which did not immediately close. At a distance of twenty miles only an occasional chimney was overturned, the walls of some brick buildings were cracked, and wooden buildings escaped without injury; the ground was not cracked, landslides were rare, and not all sleepers were wakened. At seventy-five miles the shock was observed by nearly all persons awake at the time, but there were no destructive effects; and at two hundred miles it was perceived by only a few persons.

A number of exceptions to this gradation of intensity are connected with tracts of deep alluvial soil, especially if saturated with water, and with tracts of "made ground." The great destruction in the low-lying part of San Francisco, eight miles from the fault, is directly connected with the fact that much of the ground there is artificial, the area having been reclaimed from the bay by filling in with sand and other materials. [. . .]

The most important practical results of the various earthquake studies will probably be afforded by the engineers and architects, and will lead to the construction of safer buildings in all parts of the country specially liable to earthquakes; but the geologic studies of the State Commission are not devoid of economic bearings. In the city of San Francisco and adjacent parts of the peninsula on which it stands the underlying formations include several distinct types, and the district is so generally occupied by buildings that the relations of the several formations to earthquake injury can readily be studied. Such a study is being made with care and thoroughness, and one of its results will be a map of the city showing the relation of the isoseismals, or lines marking grades of intensity, to tracts of solid rock, to tracts of dune sand in its natural position, to upland hollows partially filled by grading, and to old swamps, lagoons and tidal marshes that have been converted into dry land by extensive artificial deposits. The information contained in such a map should guide the reconstruction and future expansion of the city, not by determining the avoidance of unfavorable sites, but by showing in what areas exceptional precautions are needed, and what areas demand only ordinary precautions.

Another economic subject to which the commission may be expected to give attention is what might be called the earthquake outlook. Must the citizens of San Francisco and the bay district face the danger of experiencing within a few generations a shock equal to or even greater than the one to which they have just been subjected? Or have they earned by their recent calamity a long immunity from violent disturbance? If these questions could be answered in an authoritative way, or if a forecast could be made with a fair degree of probability, much good might result; and even if nothing more shall be possible than a cautious discussion of the data, I believe such a discussion should be undertaken and published. Of snap judgments there has been no lack, and the California press has catered to a natural desire of the commercial public for an optimistic view; but no opinion has yet been fortified by an adequate statement of the pertinent facts. Among these facts are the distribution of earthquake shocks as to locality, time and severity in California, and also in the well-studied earthquake district of Japan; the relation of the slipping that has just occurred to the geologic structure of the coast region; the relation of other fault lines to the bay district; and the relation of the recent shock to a destructive shock that occurred in 1868. If a broad and candid review of the facts shall give warrant for a forecast of practical immunity, the deep-rooted anxiety of the community will find therein a measure of relief. If a forecast of immunity shall not be warranted, the public should have the benefit of that information, to the end that it shall fully heed the counsel of those who maintain that the new city should be earthquake-proof. In any case, timidity will cause some to remove from the shaken district and will deter others who were contemplating immigration; but such considerations have only temporary influence and can not check in an important way the growth of the city. The destiny of San Francisco depends on the capacity and security of its harbor, on the wealth of the country behind it, and on its geographic relation to the commerce of the Pacific. Whatever the earthquake danger may be, it is a thing to be dealt with on the ground by skilful engineering, not avoided by flight; and the proper basis for all protective measures is the fullest possible information as to the extent and character of the danger. ■

Mary Austin

"The Temblor: A Personal Narration" (1906–7)

There are some fortunes harder to bear once they are done with than while they are doing, and there are three things that I shall never be able to abide in quietness again—the smell of burning, the creaking of house-beams in the night, and the roar of a great city going past me in the street.

Ours was a quiet neighborhood in the best times; undisturbed except by the hawker's cry or the seldom whistling hum of the wire, and in the two days following April eighteenth, it became a little lane out of Destruction. The first thing I was aware of was being wakened sharply to see my bureau lunging solemnly at me across the width of the room. It got up first on one castor and then on another, like the table at a séance, and wagged its top portentously. It was an antique pattern, tall and marble-topped, and quite heavy enough to seem for the moment sufficient cause for all the uproar. Then I remember standing in the doorway to see the great barred leaves of the entrance on the second floor part quietly as under an unseen hand, and beyond them, in the morning grayness, the rose tree and the palms replacing one another, as in a moving picture, and suddenly an eruption of nightgowned figures crying out that it was only an earthquake, but I had already made this discovery for myself as I recall trying to explain. Nobody having suffered much in our immediate vicinity, we were left free to perceive that the very instant after the quake was tempered by the half-humorous, wholly American appreciation of a thoroughly good job. Half an hour after the temblor people sitting on their doorsteps, in bathrobes and kimonos, were admitting to each other with a half twist of laughter between tremblings that it was a really creditable shake.

The appreciation of calamity widened slowly as water rays on a mantling pond. Mercifully the temblor came at an hour when families had not divided for the day, but live wires sagging across housetops were to outdo the damage of falling walls. Almost before the dust of ruined walls had ceased rising, smoke began to go up against the sun, which, by nine of the clock, showed bloodshot through it as the eye of Disaster.

It is perfectly safe to believe anything any one tells you of personal adventure; the inventive faculty does not exist which could outdo the actuality; little things prick themselves on the attention as the index of the greater horror.

I remember distinctly that in the first considered interval after the temblor, I went about and took all the flowers out of the vases to save the water that was left; and that I went longer without washing my face than I ever expect to again.

I recall the red flare of a potted geranium undisturbed on a window ledge in a wall of which the brickwork dropped outward, while the roof had gone through the flooring; and the cross-section of a lodging house parted cleanly with all the little rooms unaltered, and the halls like burrows, as if it were the home of some superior sort of insect laid open to the microscope.

South of Market, in the district known as the Mission, there were cheap man-traps folded in like pasteboard, and from these, before the rip of the flames blotted out the sound, arose the thin, long scream of mortal agony.

Down on Market Street Wednesday morning, when the smoke from the burning blocks behind began to pour through the windows we saw an Italian woman kneeling on the street corner praying quietly. Her cheap belongings were scattered beside her on the ground and the crowd trampled them; a child lay on a heap of clothes and bedding beside her, covered and very quiet. The woman opened her eyes now and then, looked at the reddening smoke and addressed herself to prayer as one sure of the stroke of fate. It was not until several days later that it occurred to me why the baby lay so quiet, and why the woman prayed instead of flying.

Not far from there, a day-old bride waited while her husband went back to the ruined hotel for some papers he had left, and the cornice fell on him; then a man who had known him, but not that he was married, came by and carried away the body and shipped it out of the city, so that for four days the bride knew not what had become of him.

There was a young man who, seeing a broken and dismantled grocery, meant no more than to save some food, for already the certainty of famine was upon the city—and was shot for looting. Then his women came and carried the body away, mother and betrothed, and laid it on the grass until space could be found for burial. They drew a handkerchief over its face, and sat quietly beside it without bitterness or weeping. It was all like this, broken bits of human tragedy, curiously unrelated, inconsequential, disrupted by the temblor, impossible to this day to gather up and compose into a proper picture.

The largeness of the event had the effect of reducing private sorrow to a mere pin prick and a point of time. Everybody tells you tales like this with more or less detail. It was reported that two blocks from us a man lay all day with a placard on his breast that he was shot for looting, and no one denied the aptness of the warning. The will of the people was toward authority, and everywhere the tread of soldiery brought a relieved sense of things orderly and secure. It was not as if the city had waited for martial law to be declared, but as if it precipitated itself into that state by instinct as its best refuge.

In the parks were the refugees huddled on the damp sod with insufficient bedding and less food and no water. They laughed. They had come out of their homes with scant possessions, often the least serviceable. They had lost business and clientage and tools, and they did not know if their friends had fared worse. Hot, stifling smoke billowed down upon them, cinders pattered like hail—and they laughed—not hysteria, but the laughter of unbroken courage.

That exodus to the park did not begin in our neighborhood until the second day; all the first day was spent in seeing such things as I relate, while confidently expecting the wind to blow the fire another way. Safe to say one-half the loss of household goods might have been averted, had not the residents been too sure of such exemption. It happened not infrequently that when a man had seen his women safe he went out to relief work and returning found smoking ashes—and the family had left no address. We were told of those who had dead in their households who took them up and fled with them to the likeliest place in the hope of burial, but before it had been accomplished were pushed forward by the flames. Yet to have taken part in that agonized race for the open was worth all it cost in goods.

Before the red night paled into murky dawn thousands of people were vomited out of the angry throat of the street far down toward Market. Even the smallest child carried something, or pushed it before him on a rocking chair, or dragged it behind him in a trunk, and the thing he carried was the index of the refugee's strongest bent. All the women saved their best hats and their babies, and, if there were no babies, some of them pushed pianos up the cement pavements.

All the faces were smutched and pallid, all the figures sloped steadily forward toward the cleared places. Behind them the expelling fire bent out over the lines of flight, the writhing smoke stooped and waved, a fine rain of cinders pattered and rustled over all the folks, and charred bits of the burning fled in the heated air and dropped among the goods. There was a strange, hot, sickish smell in the street as if it had become the hollow slot of some fiery breathing snake. I came out and stood in the pale pinkish glow and saw a man I knew hurrying down toward the gutted district, the badge of a relief committee fluttering on his coat. "Bob," I said, "it looks like the day of judgment!" He cast back at me over his shoulder unveiled disgust at the inadequacy of my terms. "Aw!" he said, "it looks like hell!"

It was a well-bred community that poured itself out into Jefferson Square, where I lay with my friend's goods, and we were packed too close for most of the minor decencies, but nobody forgot his manners. "Beg pardon!" said a man hovering over me with a 200-pound trunk. "Not at all!" I answered making myself thin for him to step over. With an "Excuse me, madam!" another, fleeing from the too-heated border of the park to its packed center, deftly up-ended a roll of bedding, turned it across the woman who lay next to me—and the woman smiled.

Right here, if you had time for it, you gripped the large, essential spirit of the West, the ability to dramatize its own activity, and, while continuing in it, to stand off and be vastly entertained by it. In spite of individual heartsinkings, the San Franciscans during the week never lost the spirited sense of being audience to their own performance. Large figures of adventure moved through the murk of those days—Denman going out with his gun and holding up express wagons with expensively saved goods, which were dumped out on sidewalks that food might be carried to unfed hundreds; Father Ramm cutting away the timbers of St. Mary's tower, while the red glow crept across the charred cross out of reach of the hose; and the humble sacrifices—the woman who shared her full breast with the child of another whose fountain had failed from weariness and fright—would that I had her name to hold in remembrance! She had stopped in the middle of a long residence hill and rested on a forsaken stoop, nourishing her child quietly, when the other woman came by panting, fainting and afraid, not of her class, nor her race, but the hungry baby yearned toward the uncovered breast—and they both of them understood that speech well enough.

Everybody tells you tales like this, more, and better. All along the fire line of Van Ness Avenue, heroic episodes transpired like groups in a frieze against the writhing background of furnace-heated flame; and, for a pediment to the frieze, rows of houseless, possessionless people wrapped in a large, impersonal appreciation of the spectacle.

From Gough Street, looking down, we saw the great tide of fire roaring in the hollow toward Russian Hill; burning so steadily for all it burned so fast that it had the effect of immense deliberation; roaring on toward miles of uninhabited dwellings so lately emptied of life that they appeared consciously to await their immolation; beyond the line of roofs, the hill, standing up darkly against the glow of other incalculable fires, the uplift of flames from viewless intricacies of destruction, sparks belching furiously intermittent like the spray of bursting seas. Low down in front ran besmirched Lilliputians training inadequate hose and creating tiny explosions of a block or so of expensive dwellings by which the rest of us were ultimately saved; and high against the tip of flames where it ran out in broken sparks, the figure of the priest chopping steadily at the tower with the constrained small movement of a mechanical toy.

Observe that a moment since I said houseless people, not homeless; for it comes to this with the bulk of San Franciscans, that they discovered the place and the spirit to be home rather than the walls and the furnishings. No matter how the insurance totals foot up, what landmarks, what treasures of art are evanished, San Francisco, *our* San Francisco is all there yet. Fast as the tall banners of smoke rose up and the flames reddened them, rose up with it something impalpable, like an exhalation. We saw it breaking up in the movements of the refugees, heard it in the tones of their voices, felt it as they wrestled in the teeth of destruction. The sharp sentences by which men called to each other to note the behavior of brick and stone dwellings contained a hint of a warning already accepted for the new building before the old had crumbled. When the heat of conflagration outran the flames and reaching over wide avenues caught high gables and crosses of church steeples, men watching them smoke and blister and crackle into flame, said shortly, "No more wooden towers for San Francisco!" and saved their breath to run with the hose.

What distinguishes the personal experience of the destruction of the gray city from all like disasters of record, is the keen appreciation of the deathlessness of the spirit of living. For the greater part of this disaster— the irreclaimable loss of goods and houses, the violent deaths—was due chiefly to man-contrivances, to the sinking of made ground, to huddled buildings cheapened by greed, to insensate clinging to the outer shells of life; the strong tug of nature was always toward the renewal of it. Births near their time came on hurriedly; children were delivered in the streets or the midst of burnings, and none the worse for the absence of conventional circumstance; marriages were made amazingly, as the disorder of the social world threw all men back severely upon its primal institutions.

After a great lapse of time, when earthquake stories had become matter for humorous reminiscence, burning blocks topics of daily news, and standing in the bread line a fixed habit—by the morning of the third day, to be exact—there arose a threat of peril greater than the thirst or famine, which all the world rose up swiftly to relieve.

Thousands of families had camped in parks not meant to be lived in, but to be looked at; lacking the most elementary means of sanitation. With the rising of the sun, a stench arose from these places and increased perceptibly; spreading with it like an exhalation, went the fear of pestilence. But this at least was a dread that every man could fight at his own camp, and the fight was the modern conviction of the relativity of sanitation to health. By mid-morning the condition of Jefferson Square was such that I should not have trusted myself to it for three hours more, but in three hours it was made safe by no more organized effort than came of the intelligent recognition of the peril. They cleaned the camp first, and organized committees of sanitation afterward.

There have been some unconsidered references of the earthquake disaster to the judgment of God; happily not much of it, but enough to make pertinent some conclusions that shaped themselves swiftly as the city fought and ran. Not to quarrel with the intelligence that reads God behind seismic disturbance, one must still note that the actual damage done by God to the city was small beside the possibilities for damage that reside in man-contrivances; for most man-made things do inherently carry the elements of their own destruction.

How much of all that happened of distress and inestimable loss could have been averted if men would live along the line of the Original Intention, with wide, clean breathing spaces and room for green growing things to push up between?

I have an indistinct impression that the calendar time spent in the city after the temblor was about ten days. I remember the night of rain, and seeing a grown man sitting on a curbstone the morning after, sobbing in the final break-down of bodily endurance. I remember too the sigh of the wind through windows of desolate walls, and the screech and clack of ruined cornices in the red noisy night, and the cheerful banging of pianos in the camps; the burials in trenches and the little, bluish, grave-long heaps of burning among the ruins of Chinatown, and the laughter that shook us as in the midst of the ashy desert we poured in dogged stream to the ferry, at a placard that in a half-burned building where activity had begun again, swung about in the wind and displayed this legend:

DON'T TALK EARTHQUAKE
TALK BUSINESS

All these things seem to have occurred within a short space of days, but when I came out at last at Berkeley—too blossomy, too full-leafed, too radiant—by this token I knew that a great hiatus had taken place. It had been long enough to forget that the smell of sun-steeped roses could be sweet. ■

John McPhee

FROM "Assembling California" in *Annals of the Former World* (1998)

In 1906, the great earthquake was an unforeseeable Act of God. Now the question was no longer *whether* a great earthquake would happen but *when*. No longer could anyone imagine that when the strain is released it is gone forever. Yet people began referring to a chimeric temblor they called "the big one," as if some disaster of unique magnitude were waiting to happen. California has not assembled on creep. Great earthquakes are all over the geology. A big one will always be in the offing. The big one is plate tectonics.

At one time and another, for the most part with [Eldridge] Moores, I have travelled the San Andreas Fault from the base of the Transverse Ranges outside Los Angeles to the rocky coast well north of San Francisco. In clear weather, a pilot with no radio and no instrumentation could easily fly those four hundred miles navigating only by the fault. The trace disappears here and again under wooded highlands, yet the San Andreas by and large is not only evident but also something to see—like the beaten track of a great migration, like a surgical scar on a belly. In the south, where State Route 14 climbs out of Palmdale on its way to Los Angeles, it cuts across the fault zone through two high roadcuts in which Pliocene sediments look like rolled-up magazines, representing not one tectonic event but a whole working series of them, exposed at the height of the action. On the geologic time scale, the zone's continual agitation has been frequent enough to be regarded as continuous, but in the here and now of human time the rift extends quietly northward through serene, appealing country: grasses rich in the fault trough, ridges intimate on the two sides—a world of tight corrals and trim post offices in towns that are named for sag ponds.

Farther north, it loses, for a while, its domestic charm. Almost all water disappears in a desert scene that, for California, is unusually placed. The Carrizo Plain, only forty miles into the Coast Ranges from the ocean at Santa Barbara, closely resembles a south Nevada basin. Between the

Caliente Range and the Temblor Range, the San Andreas Fault runs up this flat, unvegetated, linear valley in full exposure of its benches and scarps, its elongate grabens and beheaded channels, its desiccated sag ponds and dry deflected streams. From the air the fault trace is keloid, virtually organic in its insistence and creep—north forty degrees west. On the ground, standing on desert pavement in a hot dry wind, you are literally entrenched in the plate boundary. You can see nearly four thousand years of motion in the bed of a single intermittent stream. The bouldery brook, bone dry, is fairly straight as it comes down the slopes of the Temblor Range, but the San Andreas has thrown up a shutter ridge—a sort of sliding wall—that blocks its path. The stream turns ninety degrees right and explores the plate boundary for four hundred and fifty feet before it discovers its offset bed, into which it turns west among cobbles and boulders of Salinian granite. ▪

Eliza Ruhamah Scidmore

FROM "The Recent Earthquake Wave on the Coast of Japan"
(originally published in *National Geographic Magazine,*
September 1896)

On the evening of June 15, 1896, the northeast coast of Hondo, the main island of Japan, was struck by a great earthquake wave (*tsunami*), which was more destructive of life and property than any earthquake convulsion of this century in that empire. The whole coastline of the San-Riku, the three provinces of Rikuzen, Rikuchu, and Rikuoku, from the island of Kinkwazan, 38° 20' north, northward for 175 miles, was laid waste by a great wave moving from the east and south, that struck the San-Riku coast and in a trice obliterated towns and villages, killed 26,975 people out of the original population, and grievously wounded the 5,390 survivors. It washed away and wrecked 9,313 houses, stranded some 300 larger craft—steamers, schooners, and junks—and crushed or carried away 10,000 fishing boats, destroying property to the value of six million yen. Offshore rocks were broken, overturned, or moved hundreds of yards, shallows and bars were formed, and in some localities the entire shoreline was changed.

A high mountain range bars communication with the trunk railway line of the island, and this picturesque, fiord-cut coast is so remote and so isolated that only two foreigners had been seen in the region in ten years, with the exception of the French mission priest, Father Raspail, who lost his life in the flood. With telegraph offices, instruments, and operators carried away, word came slowly to Tokyo, and with 50 to 100 miles of mountain roads between the nearest railway station and the seacoast, aid was long in reaching the wretched survivors. The first to reach the scene of the disaster was an American missionary, the Rev. Rothesay Miller, who made the usual three days' trip over the mountains in less than a day and a half on his American bicycle.

There were old traditions of such earthquake waves on this coast but the water barometer gave no warning, no indication of any unusual conditions on June 15, and the occurrence of thirteen light earthquake shocks during the day excited no comment. The villagers on that remote coast adhered to the old calendar in observing their local fêtes and holidays, and on that fifth day of the fifth moon had been celebrating the Girls' Festival. Rain had driven them indoors with the darkness, and nearly all were in their houses at eight o'clock, when, with a rumbling as of heavy cannonading out at sea, a roar, and the crash and crackling of timbers, they were suddenly engulfed in the swirling waters. Only a few survivors on all that length of coast saw the advancing wave, one of them telling that the water first receded some 600 yards from ghastly white sands and then the Wave stood like a black wall 80 feet in height, with phosphorescent lights gleaming along its crest. Others, hearing a distant roar, saw a dark shadow seaward and ran to high ground, crying *"Tsunami! Tsunami!"* Some who ran to the upper stories of their houses for safety were drowned, crushed, or imprisoned there, only a few breaking through the roofs or escaping after the water subsided.

Shallow water and outlying islands broke the force of the wave in some places, and in long, narrow inlets or fiords the giant roller was broken into two, three, and even six waves, that crashed upon the shore in succession. Ships and junks were carried one and two miles inland, left on hilltops, treetops, and in the midst of fields uninjured or mixed up with the ruins of houses, the rest engulfed or swept seaward. Many survivors, swept away by the waters, were cast ashore on outlying islands or seized bits of wreckage and kept afloat. On the open coast the wave came and withdrew within five minutes, while in long inlets the waters boiled and surged for nearly a half hour before subsiding. The best swimmers were helpless in the first swirl of water, and nearly all the bodies recovered were frightfully battered and mutilated, rolled over and driven against rocks, struck by and crushed between timbers. The force of the wave cut down groves of large pine trees to short stumps, snapped thick granite posts of temple gates and carried the stone cross-beams 300 yards away. Many people were lost through running back to save others or to save their valuables. ■

4 Volcanoes and Eruptions

The pilot flying towards the Straits of Magellan sees
below him, a little to the south of the Gallegos River,
an ancient lava flow, an erupted waste of a thickness
of sixty feet that crushes down the plain on which it
has congealed. Farther south he meets a second flow,
then a third; and thereafter every hump on the globe,
every mound a few hundred feet high, carries a crater
in its flank. No Vesuvius rises up to reign in the clouds;
merely, flat on the plain, a succession of gaping howitzer
mouths.

This day, as I fly, the lava world is calm. There
is something surprising in the tranquillity of this
deserted landscape where once a thousand volcanoes
boomed to each other in their great subterranean
organs and spat forth their fire. I fly over a world
mute and abandoned, strewn with black glaciers.

—Antoine de Saint-Exupéry,
Wind, Sand, and Stars
translated from the French by Lewis Galantière

VOLCANIC ERUPTIONS AND VOLCANOES ARE DRAMATIC SURFACE EXPRES-
sions of Earth's internal heat engine. Igneous rocks give us a window to the
remote interior world where magmas form. The term *igneous,* from the Latin
ignis for fire describes rocks that form by cooling and solidification of molten
rock (magma) from deep inside Earth's interior. These rocks are further subdi-
vided as *plutonic* (from the Greek god Pluto, god of the underworld), which
solidify beneath Earth's surface, and *volcanic* (from the Roman god Vulcan,
god of fire [Hephaistos in Greek]), which pour out on Earth's surface.

As with earthquakes, the location of the more than five hundred active volcanoes is not random. Among the planet's most striking landscape features are
the chains of volcanoes that parallel convergent plate boundaries—80 percent
of active volcanoes occur here. As one converging plate is forced to descend
into Earth's interior beneath the other plate, some of both plates' rocks melt and
cause active volcanoes to form, such as those outlining much of the so-called
Ring of Fire around the Pacific Ocean. Volcanoes in Japan, Indonesia, the
Philippines, the Aleutian Islands, and large parts of the west coast of North,
Central, and South America include some of the most active and best known
in the world. Eruptions of volcanoes such as Indonesia's Tambora in 1815
(which caused the "Year without a Summer" in 1816 in Europe and North
America) and Krakatau in 1883 (one of the most destructive explosive eruptions ever described), and the relatively small, but widely observed and
reported, 1980 eruption of Mount St. Helens are well-known examples.

The landscapes and mountains that make up the Ring of Fire, and other volcanic regions have inspired description, contemplation, and analysis. The
pieces included in this section demonstrate great breadth in content and purpose, from scientific observations to personal reflection and poetry.

Several selections consider volcanic elements of the Ring of Fire. We have
included works by contemporary American writers John Calderazzo and
Ursula K. Le Guin on, respectively, the eruptions of Alaska's Redoubt volcano
in 1989 and the Cascades' Mount St. Helens in 1980; and Charles Darwin on
his observations in 1835, while on the *HMS Beagle* voyage, of the nature, tim-
ing, and extent of eruptions in the Chilean Andes. Each work conveys a sense
of awe and regard for the scale and power of these lands born of fire.

Although most volcanic activity is concentrated at convergent and divergent plate boundaries, volcanoes within plates can also produce enormous eruptions. Geoscientists currently think that mid-plate volcanoes develop over *hotspots,* which may be surface manifestations of jets or plumes of hot material that rise from deep within Earth's mantle, drill through the crust, and emerge as volcanic centers. As a plate rides over a hotspot, the location of current volcanic activity shifts with plate movement. One of the best-known examples is the island chain that begins with the active volcanoes on Hawai'i and continues northwestward as a string of progressively older, extinct, eroded, and ultimately submerged volcanic peaks called the Emperor Seamounts. Rising ten kilometers above the ocean floor to its summit, Mauna Loa on the island of Hawai'i is the largest volcano and the tallest mountain on Earth. And Loihi, the young submarine volcano to the southeast, may become a new Hawai'ian island in the future. Poet and teacher Garrett Hongo, who was born in Volcano, Hawai'i, in 1951, recounts his rediscovery of that land in his memoir, *Volcano* (1995). In "Fire in the Night," James D. Houston (1933–) describes his experience walking on lava with U.S. Geological Survey geologist Jack Lockwood. The author of several works of fiction and nonfiction, Houston coauthored with his wife, Jeanne Wakatsuki Houston, *Farewell to Manzanar,* an account of her family's internment at Manzanar, California, during the Second World War.

The island of Iceland is thought to represent the product of eruptions from a hotspot located at a divergent plate margin, the Mid-Atlantic Ridge. A catastrophic eruption occurred there in 1783 along one of the few exposed segments of the mid-ocean ridge when a fissure tens of kilometers long erupted, killing approximately one-fifth of Iceland's population. Eruptions continue to occur along opening cracks as Iceland is pulled apart by the creeping divergence of the Eurasian plate eastward and the North American plate westward at the rate of approximately two centimeters a year. In a selection from *The Control of Nature* (1989), John McPhee writes of Iceland's campaign to cool and stop the advance of lava pouring from a fissure eruption on the island of Heimaey in 1973.

In September and October 1835, the *HMS Beagle* reached the volcanic Galápagos Islands, possibly formed by a hotspot lying far west of the coast of Ecuador along the equator. Charles Darwin published his systematic observations of the natural history and geology of the islands in *Voyage of the Beagle*. In his description, Darwin hints at his work yet to come on the origin of species by noting that the "natural history of these islands is eminently curious, and well deserves attention," and that "most of the organic productions are aboriginal creations, found nowhere else; there is even a difference between the inhabitants of the different islands." Contemporary American writer Annie Dillard describes a more recent experience on the Galápagos Islands in an excerpt from *Teaching a Stone to Talk* (1982).

We also include a letter written to the Roman historian Tacitus by Pliny the Younger, describing the death of his uncle Pliny the Elder during the A.D. 79 eruption of Mount Vesuvius. At the time of this eruption, which destroyed Pompeii and Herculaneum, Pliny the Elder commanded the Roman fleet across the Bay of Naples at Misenum.

From *Craters of Fire* (1952), French volcanologist and filmmaker Haroun Tazieff (1914–98) recalls a not-so-sensible walk around the rim of an active volcano.

John Calderazzo

FROM *Rising Fire: Volcanoes and Our Inner Lives* (2004)

A surprising and potentially terrifying consequence of volcanic eruptions has come to light only in the last few decades. This consequence had to wait for the development of a seemingly benign piece of technology: the jet engine. But what a volcano can do to that engine is not benign at all.

On December 15, 1989, an almost brand-new Boeing 747 jumbo jet took off from the Netherlands, bound for Tokyo. It didn't fly east from Amsterdam as you might imagine it would. It headed north, over the top of the world, following an arc over the frozen sea of the North Pole and then down toward Anchorage, Alaska, where it was scheduled to refuel and let off passengers before completing its run to Tokyo.

Look at a world map, a flat representation of a sphere, and this route at first seems to make little sense. But run your finger over a globe of the earth, and three-dimensional logic quickly shows that the shortest distance between Chicago and Hong Kong, or New York and Beijing, or Amsterdam and Tokyo (or at least the shortest route that also skirts spookily chaotic Russian airspace), takes you over or near Alaska. These are among the "great-circle routes" followed by thousands of international jetliners every year.

Add to them the local and regional flights that crisscross a mostly roadless state that's more than twice as large as Texas, and you have a lot of aircraft plying Alaskan skies. Puttering bush planes. Helicopters. Air Force jets and C-130s. Fed-Ex, UPS, and other freight flights. Anchorage International Airport handles more international cargo, in dollar value, than any other U.S. airport. In all, at least 6,000 planes a day travel the Alaskan skies.

The Amsterdam jet, KLM Flight 867, entered northern Alaskan airspace around 11 A.M. The sun had just risen, slightly, over the absurdly brief Arctic day. There were 231 passengers on board, plus a crew of thirteen.

The jet was also beginning to pass over some of the most deeply unstable landscape on the planet. Thanks to the fractured geometry of plate tectonics, the northward-crawling Pacific oceanic plate creeps in a downward curve two or three inches per year beneath the North American continental plate. As the forward edge of this ocean-bottom crust grinds fifty miles or so below the Alaskan mainland, stresses and strains build up and are released in sometimes subtle, sometimes furious jolts, endless palsies in the earth.

The converging plates have crumpled up the Alaskan landscape into spectacular mountains. The plate movement has also helped give birth to more than forty of the fifty-four U.S. volcanoes that have been active in historic time. That's about eight percent of all active above-water volcanoes on earth. Most of them are strung out, like beads on a necklace, along the 1,550-mile-long Aleutian Arc that runs west from Anchorage to Russia's Kamchatka Peninsula, which itself contains dozens of active volcanoes. The raw and windswept chain forms the northernmost portion of the Pacific Ring of Fire.

One of the most prominent mountains in this long arc is 10,197-foot-tall Redoubt volcano, a thickset triangle of ice and snow that hulks over Cook Inlet, in Lake Clark National Park, about 100 miles southwest of Anchorage. The summit crater of Redoubt has been fairly active over the years. Up until 1967, it had gone off, moderately, six times during the twentieth century.

An hour or so before Flight 867 reached Alaskan airspace, Redoubt, which had been rumbling seriously all day and threatening to blow for the first time in more than twenty years, erupted explosively. It roared for forty minutes, sending up an ash-rich plume that quickly grew to resemble a mushroom cloud like the ones from those South Pacific atomic bomb tests. Soon the cloud had climbed to a height of seven-and-a-half miles above sea level. At that altitude, which is higher than the summit of Mount Everest and just past the point where the troposphere thins out into the stratosphere, hurricane-speed winds are routine. Consequently, the top of the ash plume, ever-widening, began to race away from the volcano.

The giant cloud bolted to the northeast. Scalding hot subterranean rock had just turned into "weather."

A modern jetliner is a wonder of technology. But all instruments have their limits, and the radar system on an ordinary passenger jet cannot distinguish an ash cloud from an ordinary thunderhead. Before take-off, 867's flight crew had been briefed about the potential of a Redoubt eruption—although the exact time and size would be only the wildest of guesses—so they'd taken on 5,000 extra gallons of fuel in case they had to divert.

As they eased into their long descent to Anchorage, the pilots saw nothing in particular to divert from. They'd been informed by the Alaska Volcano Observatory, then in its first year of operation, that Redoubt had indeed recently erupted. Yet they were still more than 200 miles from the volcano. And if the cloud layer the pilots were bearing down on looked a bit browner than normal—a little bruised, maybe—they apparently felt no alarm. They barreled into it at about 500 mph.

There was instant darkness. Darkness, and then, for a few magical seconds, lighted particles zooming past the cockpit like a thousand fireflies, a million fireflies. It looked unworldly; it was a vision of the universe just seconds after the Big Bang—an incredible blast of stars. And then after a few seconds the crew saw nothing at all, for their acrylic windshields, the leading edges of the airliner, had been abraded as though in a sand storm. Fine brown dust began to swirl in the cockpit.

In the passenger cabin, the lights went out and a powerful smell of sulfur came through the vents. Passengers pulled out handkerchiefs and held them to their noses and mouths. Some of them saw coronas or eerie lights playing about the wings and engines. Generated by the supercharged ash particles grinding all around the plane, these electrical discharges were almost certainly St. Elmo's Fire, which during the Middle Ages had been named after the patron saint of sailors. Seafarers from Columbus to Magellan to the fictional, doomed crew of *Moby Dick*'s Pequod reported seeing the saint's "holy body" glow briefly as an electrical discharge on the tips of their vessel's masts or yardarms. Sometimes the fire danced on harpoon points or the heads of spears. Many regarded it as a portent of bad weather.

By now, KLM 867 was definitely plowing through bad weather. The pilots noticed a "Cargo Fire" warning light blink on, but thinking that it was caused by what they were quickly realizing was volcanic ash, they decided to ignore it. Besides, there were more urgent matters to attend to. The gloom in the cockpit had become so thick that the pilot could find her co-pilot only by reaching out and touching his shoulder. The cockpit crew strapped on oxygen masks. Once she could breathe normally again, the pilot veered left and started a full thrust climb to escape the ash as quickly as possible. At first the jet performed normally. But the more it climbed, the more it began to lose air speed.

"Ash" is a funny word. Unless you're a volcanologist, it usually brings to mind the gray, feathery remains of a woodstove, or dying campfire coals. But nothing soft or fluffy blasts out of the throat of a volcano. What shoots out is a magmatic mix of minerals and glass shards, fragmented and powered by violently expanding gases and steam. In other words, magma that's been blown to smithereens, blown into billions of pieces of fast-cooling, sharp-edged grit often smaller than the period at the end of this sentence.

Soon these particles were scraping the leading edges of the jet's wings and tail rudder. The finer pieces worked their way into the hydraulic system and fuel sump. They even seeped into the engine oil. More than seven tons of air per minute were being sucked through each of 867's engines, and the ash that now clogged that air began to gnaw away at the turbofan blades. In the combustion chambers, which operate at temperatures slightly above the melting point of ash, the particles cooked, then solidified as glass and coated the turbine blades in a ceramic skin. This fooled the jet's safety sensors into thinking that the engines were overheating. Then the sensors shut them down.

One minute into 867's diversionary climb, all four engines went dead. The jet was five miles above sea level, and three or so above the craggy, icebound peaks of the Alaska Range.

"ANCHORAGE," the pilot radioed, "WE ARE IN A FALL!!"

When its engines fail, an airliner traveling at high speed doesn't just drop like a rock. It glides, at least for a while. KLM 867 had been climbing at a rate of 1,500 feet per minute. Now, in a great silence, it gradually leveled off and began to arrow slightly down, as though an enormous invisible rock had come to sit on its nose. The rock was gravity.

The pilots tried to restart the engines. Nothing. They tried again. Nothing. In a powerless glide, the jet dropped faster. Soon, pencils and pens began to float about the passenger cabin. So did thick novels, bags of peanuts, plastic cups, playing cards, pillows, blankets, all of them floating in the dim, acrid air.

For five minutes the plane fell in silence, or at least a mechanical silence, because by now many of the passengers were screaming or retching or holding on to each other and praying. The pilots tried again and again to restart the engines. While they worked, the plane dropped out of the bottom of the ash cloud. The passengers could now see for themselves how close they were to the jagged peaks of the icebound mountains. On the eighth or ninth restart try, the now-cold glass that coated the turbofan blades in one of the left-wing engines fractured, and the blades bit into air. A moment later the other left-wing engine roared back to life. Three minutes later, the right-wing engines followed. With a few thousand feet to spare, the pilot managed to level off, then turn for Anchorage.

They landed without further incident, some in the cockpit crew leaning far to the side and squinting out of the slightly less frosted edges of the windshield. None of the passengers had been injured, at least not physically, but the engines were ruined. The jet suffered $80 million in damage and was grounded for months.

Meanwhile, the ash cloud kept moving. Although it didn't clog their engines, the ash scraped up two other nearby jets. As pilot and air control chatter filled the skies, rippling out over much of the northern hemisphere, the wandering cloud scared off many more aircraft. Growing ever more diffuse, the ash was pushed south by high wind currents toward Mexico. Two days later, as it passed over the skies of El Paso, Texas, it still carried enough particles to scratch up the wings of a Navy DC-9 that had no idea where the floating grit had come from. El Paso was 3,000 miles from the broken top of Redoubt volcano.

It is sobering to think that all of this middle- and long-distance chaos (and potential disaster) on the ground and high in the air was set in motion by a very ordinary-sized eruption. Compared to some of the past century's larger or more spectacular eruptions, such as Pinatubo in the Philippines or Mount St. Helens, Redoubt volcano went off like a popgun. And it still almost dragged down a jetliner 200 miles away. ▪

Garrett Hongo

FROM *Volcano: A Memoir of Hawai'i* (1995)

Ūgetsu

I was born in the village of Volcano, in the back room of the kitchen of a general store my grandfather built on Volcano Road, twenty-nine miles from Hilo on the Big Island of Hawai'i, the newest in the mid-Pacific chain its first human settlers called *Hava-iti,* a name which means, in their ritualized language, "the Realm of the Dead" or, more simply, "Paradise." Volcano is a settlement in the fern and *'ōhi'a* rain forests bordering two active volcanoes—Kīlauea, a huge caldera and chain of craters leading eastward to the sea, and Mauna Loa, a magnificent earthen shield that places the entire landscape under its shadow, gathering its republic of rain clouds by midafternoon every day, and, at thirteen thousand feet above sea level, the largest volcanic mass in the world. I could have grown up here, in this natural garden and amphitheater for fire-fountains and rain, except that there was some family quarrel which made my parents leave with me when I was eight months old. We ran off first to O'ahu and my mother's family, and then, finally, to Los Angeles in California, with its foreign culture and ways of anonymity. I'd grown up there, cipherously, a child of immigrants to the Mainland, and, for half a lifetime, I knew nothing of the family history or the volcanoes I was born under. I was past thirty when I first returned, and the place astonished me.

Volcano is a big chunk of the sublime I'd been born to—the craters and ancient fire pit and huge black seas of hardened lava, the rain forest lush with all varieties of ferns, orchids, exotic gingers, and wild lilies, the constant rain and sun-showers all dazzled me, exalted me. And, at nearly four thousand feet, its climate was strangely *cool,* the sweltering tropics several climatic zones below, so that temperatures stayed in the high sixties during the day, sometimes dropping to the low forties by early morning. While outdoors, the villagers all wore flannels and sweatshirts and corduroy trousers—mountaineering clothes. During that first visit, Volcano seemed to me an exotic and slightly miraculous place that I associated with the

Macondo of García Márquez, the Macchu Picchu of Pablo Neruda. There was something magical about it—a purgatorial mount in the middle of the southern ocean—and there was something of it native to me, an insinuation of secret and violent origins and an aboriginal past.

[. . .]

Eruption

One night, around the third week we'd spent in the cabin, I was annoyed awake by a rumbling I sensed underneath the *futon*. I heard a whining moan like water rushing through a big pipe, a groaning like a big bus cranking through a gearshift or going uphill. It made the floorboards of the cabin shake a little. It made the windows rattle. It was sometime past twelve o'clock and it should have been nearly completely dark, but as I popped up, hugging the bedcovers against the chill of night, I saw the sky behind the back windows of the cabin illuminated with an infernal, fiery red light that cast the stand of forest trees nearby into silhouette, black and skeletal against its glowing. The sky had gone entirely red outside except for the stark fringing of *ʻōhiʻa* trees. Kīlauea was erupting full force from a vent about nine miles away from us.

In the sleepy fog of my mind, I thought, *Oh, an eruption*, and I laid myself back down, trying to fall back asleep. I guessed the vibrating to be lava moving through the rift zone, its huge earthen conduit miles below us. I was happy in a dumb sort of way. I thought, *Oh, an eruption, and I'm here for it.*

Geologists had been expecting a new outbreak. I'd been reading the papers and hearing news bulletins and volcano updates on the car radio. There had been reports of seismic activity coming from Puʻu ʻŌʻō, a new vent site about nine miles from the summit of Kīlauea and about a quarter mile from the site of the last outbreak at Kamoamoa. Lavas from a flow emanating from Puʻu ʻŌʻō had oozed along the line of cliffs that had once made up the coast of this island some thousands of years ago, and ran down to burn a few homes near the oceanside village of Kalapana. Pele, the Hawaiian goddess of volcanic creation, had decreased her activity somewhat after that, restricting herself to fuming, steaming, and a little shaking at a spot along the East Rift Zone on old Campbell Estate land just outside the boundary to the National Park.

The East Rift was a vague line of recent activity running in a northeasterly direction from the summit of Kīlauea near Volcano Village down to the sea near the village of Kapoho. It was as if a huge underground conduit had been laid from Kīlauea Summit (where most of the nineteenth-century activity had been centered in the lava lake of Halemaʻumaʻu) easterly down to the sea. It was believed that magma came up below Kīlauea and filled a gigantic storage chamber under it about a mile deep, then, once filled, it ran out along this East Rift Zone until it simply overstrained the leaky plumbing and burst out in red blossoms of molten rock. The Kapoho eruption in 1960 had happened this way—a fountain of fluid *pāhoehoe* first breaking out in a papaya patch, rupturing the earth and sending up small, attractive fountains; then it erupted in a huge and terrifying curtain of fire a hundred feet high and forty yards long behind the village stores; finally, it evolved into a huge cinder and spatter cone that sent flows of *ʻaʻā*, that clinkery rubbled lava, enough to bury the entire village and surrounding farms. Kapoho—a cluster of houses, agricultural buildings, and a string of shops along the highway—was entirely lost.

One of my uncles had land out there then, planted in vanda and dendrobium orchids—the fleshy purple-and-yellow ones—and my eldest cousin remembers running through the orchid fields, grabbing flowers, shoving them into buckets in the back of a running jeep, and smelling the sulphur fumes of the flow heading his way. All the recent activity had been along this line, from Kīlauea Iki in 1959 and Kapoho in 1960 to Mauna Ulu (the Darth Vader of all spatter cones) from 1929 to 1974. Since January of 1983, the eruption had been at Puʻu ʻŌʻō, another site along the East Rift Zone, and episodes of fountaining had been occurring there about once a month. I knew this from having read up a little, from having had conversations with relatives, from remembering stories from my own childhood. I'd never witnessed an eruption, but I vaguely knew how it would go, or so I thought. In my drowsiness, I wanted to sleep a little more.

My wife, sitting up and gazing at the red glow in the windows, was aghast. She saw my sleepy nonchalance as pure foolishness. She thought I was irresponsible.

"I'm calling the neighbors," Cynthia said, prodding me with her knee.

My wife, from Oregon, and I, from Hawai'i, had very different responses to volcanoes. Her knowledge came from the eruption of Mount St. Helens near Portland in 1980, an explosive and deadly, bomblike blast caused by the buildup of energy released in the collision between two massive geologic formations—the Pacific and the North American plates. Mount St. Helens blew the top off itself, knocked down forests, vaporized lakes and leveled towns. It killed. But, from the reading that I'd done, my idea of a volcano was Kīlauea—a beautiful, almost continuous flowering of volcanic energy that, over eons, had slowly built this chain of islands. I'd read that eruptions here were fountains, lava displays, rivers of heaven. They were spectacles and illustrations of the world's splendor. And, given a good night's sleep, I felt completely ready to cherish them.

"Look," I said, calling after her, "if you have to call someone, why don't you call the Volcano Observatory? They'll have the news and you can ask them any questions."

The Hawaiian Volcano Observatory was a field station of the United States Geological Survey built during the early part of this century to monitor and study the activity of Kīlauea. We'd been to a viewpoint adjacent to the observatory once. It was on a high bluff with a spectacular view of Halema'uma'u, the remnant of the lava lake, and the old, gigantic caldera of Kīlauea. I noticed then that there were geologists inside one of the buildings who seemed to be on observational duty around the clock.

She phoned through, getting a technician on the line right away. "What's going on?" she asked. The technician, a local, explained that this was the nineteenth episode of the current eruptive phase of Kīlauea that began in 1983, that this was an episode of "high fountaining—to one t'ousan' feet at leas'," and that lava was geysering out of Pu'u 'Ō'ō, the main vent site just upslope from the prehistoric cone of Kamoamoa. His explanation, Cynthia told me later, was fairly technical, his delivery almost deadpan, except that he seemed to be suppressing excitement, as if he were speaking to her while staring through binoculars at the eruption. I imagined him to be a local guy dressed in rubber beach sandals, jeans, and a T-shirt while manning the hot line, stirring a cup of coffee with a wooden stirstick, checking seismographs and jotting measurements, spotting the eruption through binoculars fixed on a metal stand, holding conversations with Civil Defense and USGS headquarters, and talking to my wife all at the same time.

"Where should we go?" Cynthia asked finally, getting to her point.

"Where are you?" the technician asked.

"In Mauna Loa Estates, twelve streets in from the twenty-sixty-mile marker," she said, citing our coordinates.

"Oh," he said. "You're only a few miles from the vent. All you have to do is get in your car, drive out to Volcano Highway, go down *makai* [seaward] about five miles to Glenwood. There's a turnout there across the highway from Hirano Store. Drive up to the horsegate and park. Guarantee you can see it real good from there."

"See it? I don't want to *see* it," she exclaimed, "I can *already* see it. From my window. The sky's all red. All the windows are red. The forest is *glowing*. What about the danger? Where should I take my family? I want to get them *away* from it!"

The technician laughed. He asked her where she was from. She told him. He laughed again. "There's no *danger*," he said. "I thought you were calling to find out the best place to *see* it!"

"Oh," Cynthia said, and she hung up, a little mad and a lot relieved. "I guess it's no big deal," she said, turning to me. We hugged each other.

"Let's go look at it," she said.

I drove us according to the technician's directions, then downslope along the highway to the turnout in Glenwood, pulling up alongside the horsegate he spoke of. We looked out from the car and saw only a veil of overcast tinged red from the erupting volcano. It was vapor from the eruption that made the clouds that hid its light.

I drove us still farther downhill, hoping for another, clearer vantage point, a little turnout along the highway where a crowd of locals might be gathering to witness the emergence of Pele. At Kea'au, a crossroads town only a thousand feet from sea level, we spotted a car pulled over near a power-switching station, and a local-looking man, heavyset, bearded, dressed in shorts and a flannel shirt, leaning over a guardrailing, was gazing fixedly over fields of sugarcane at a point nearly on the horizon. I parked the car, we got Alexander out of his baby seat, and we walked over to where the man leaned against the metal railing.

Nearly back to sea level, the air there was much warmer than in Volcano, the obscuring fog high above us. We were on a slight overlook above a field of lands cultivated in orchids, black plastic and white cotton sheets of curved awning and canopy in neat rows below us. Beyond this was old caneland, abandoned to flowering, and stretching out for acres and acres in the distance on the long lava plain of Kīlauea. The man gestured with a lift of his chin, we turned to look, and we saw a thin wire of red fire lifting itself from under white and graying clouds out almost on the edge of all that we could see. Far away, from where we could make out the bare outline of a small cone, there was the rose-colored stem of light that illuminated a plume of white vapor billowing out into a sea of cloud cover spreading over the land. This was the eruption, the fire-fountain of lava, a tiny jewel of a glimpse into dread and delight.

The man was friendly. He explained he'd been watching the eruption from there for nearly an hour, how the fountain had been higher earlier on, how there had been less cloudiness, wider spreading of the lava as it spumed out from the vent. We stayed there only a short while, though, since there wasn't that much to see, and it seemed to me as if we were breaking in on the man's peace, his meditation. We drove back home, no longer excited, but feeling something else, all of our foolishness gone. ■

John McPhee

FROM "Cooling the Lava" in *The Control of Nature* (1989)

Cooling the lava was Thorbjorn's idea. He meant to stop the lava. That such a feat had not been tried, let alone accomplished, in the known history of the world did not burden Thorbjorn, who had reason to believe it could be done.

His full name is Thorbjorn Sigurgeirsson. If you look for him in Simaskra, the Iceland telephone directory, you look under "Thorbjorn." You look under "Sigurdur" for Sigurdur Jonsson. You look under "Magnus" for Magnus Magnusson. I had occasion recently to call all three of these, and finding them in Simaskra was not for me a simple task. It doesn't matter that Sigurdur is a harbor manager's deputy, or that Magnus is a postal director, or that Thorbjorn is a physicist trained in Copenhagen by Niels Bohr. Like the Prime Minister, like the President—like all people in Iceland—Thorbjorn is known by his first name. Sigurgeir, of course, was his father.

To Thorbjorn's idea skepticism was the primary response. The skeptics included Magnus Magnusson, Sigurdur Jonsson, Valdimar Jonsson, Thorleifur Einarsson, Gudmundur Karlsson, and almost everybody else in Iceland. Red-hot lava—moving with the inexorability of tide—was threatening a town and a harbor on an offshore island. The vent was a nascent volcano. As the entire nation watched on television, a small crew with fire hoses squirted the front of the lava, producing billows of steam. This was in February, 1973. Quickly, the cooling of the lava became a national joke. The people called the action pissa a hraunid, in which "a hraunid" meant "on the lava."

There may even have been smiles in the National Emergency Operation Center, in Reykjavik, where the Civil Defense Council was watching Thorbjorn's experiments. The National Emergency Operation Center has grown since then but is still concentrated around a command-post bomb shelter built as Iceland's cerebral cortex in the event of a nuclear war. The Civil Defense Council had been established in 1962, and was scarcely a year old when—sixty-five miles southeast of the capital—the ocean began to boil. The Council became preoccupied with forces greater than bombs. Red lava appeared in the Atlantic swells and, layer upon layer, emerged as an island that was soon the second largest of a fleet of islands collectively known as Vestmannaeyjar. The fresh lavas of Surtsey, as the new one was called, hissed into the ocean for three and a half years. In the same period, unusual masses of drift ice blockaded various harbors, causing wide destruction; and the charter purpose of the Civil Defense Council was broadened to include acts of God. Behind thick concrete walls and steel anti-radiation doors, in a space where the air can be pressurized to keep anything noxious from seeping in, Iceland directs the war against nature. "War" is the word often used, especially with reference to the campaign begun after the twenty-third of January, 1973, when a fissure suddenly opened in the outskirts of a community of five thousand people and a curtain of lava five hundred feet high and a mile long fountained into the sky above Heimaey, the largest island of Vestmannaeyjar. All these years later, the communication room of the National Emergency Operation Center, in Reykjavik, is decorated with a relief model of Heimaey and eight large photographs, ruddy with violence, made in that place and time, for, notwithstanding the frequency of eruptions in various parts of Iceland and of natural disasters in many forms (sea surges, earthquakes, glacial bursts), Heimaey is the most signal battle ever waged by the Civil Defense.

Heimaey is pronounced "hay may." Vestmannaeyjar is more or less pronounced "vestman air." The town on Heimaey is the only place in the archipelago inhabited by human beings, and it has no special name. It is referred to simply as Vestmannaeyjar or, less frequently, as Heimaey, which means "home island." In area, the home island is so small that it approximates Manhattan south of the Empire State Building. The volume of material that came pouring out on Heimaey in 1973 would be enough to envelop New York's entire financial district, with only the tops of the World Trade Center sticking out like ski huts. The image is not as outlandish as it seems. A few miles west of Manhattan, the high ground of Montclair—of Glen Ridge, Great Notch, and Mountainside—is the product of a similar fissure eruption. In Vestmannaeyjar, the nub of the crisis was simple and economic. Proportionally, Heimaey was more valuable to Iceland than downtown New York is to us. Vestmannaeyjar, with two and a half per cent of Iceland's people, was producing about a twelfth of Iceland's export income. Vestmannaeyjar was a place to catch fish. Iceland's exports were three-quarters fish. As a result of Heimaey's natural protection, it was Iceland's single most important fishing center. Not merely was Heimaey's harbor the best harbor along the three hundred miles of Iceland's south coast; it was the only one. That is why this eruption, as someone said, "just came like a thundering cloud over the whole nation." That is why there was so much alarm when a generally slow and viscous, magisterial lava—dark-shelled by day and a craquelure of red and black at night—began to move not only toward sections of the town but, more important, toward the entrance to the harbor.

The cooling was at first confined to the lava front. Men stood on cold ground before the flowing rock and watered it like a garden. Its Fahrenheit temperature was around two thousand degrees. The water would reduce the heat locally by a factor of four, creating a wall of chilled lava to dam the flow behind. Soon it became apparent that the wall would have to be a great deal thicker than hoses could ever make it from positions ahead of the flow. The lava should be cooled not so much by the edge as by the acre, and that called not only for more pumps but also for the deployment of matériel and personnel up on top of the advancing flow.

It was astonishing to see what an essentially liquid body of rock would carry on its surface. As lava moves, under the air, it develops a skin of glass that is broken and rebroken by the motion of the liquid below, so that it clinks and tinkles, and crackles like a campfire, which, in a fantastic sense, it resembles. In Hawaii, I have been close enough to the flowing lava of Kilauea to ladle some out and carry off a new rock, but when I visited Vestmannaeyjar the eruption had been over for a decade and a half, and, needless to say, no lava was moving. Steam was still rising from the lava field, though, and grew to heavy mist under rain. If you reached down and put a hand on the ground—on loose ash—the new surface felt cool. If you rubbed away as little as a third of an inch, the ground was so hot you had to pull back your hand. The ash was such an effective insulator—not to mention the layers of solidified basalt underneath it—that the interior of the 1973 flow was still molten. During the eruption, when the pumping crews first tried to get up onto the lava they found that a crust as thin as two inches was enough to support a person and also provide insulation from the heat—just a couple of inches of hard rock resting like pond ice upon the molten fathoms. As the crews hauled and heaved at hoses, nozzle tripods, and sections of pipe, they learned that it was best not to stand still. Often, they marched in place. Even so, their boots sometimes burst into flame.

It was soon apparent that bulldozers were required on the lava—to flatten the apalhraun (the jagged surface glass, known elsewhere in the world by the Hawaiian word "aa"), to make roads for the pumping crews, to move some very heavy pipe. With the eruption going at full volume, bulldozers crawled up onto the flow. About a foot and a half of crust was enough to support them—to keep them from plunging through. The bulldozers worked in dense warm fog—the remains of the seawater pumped on the lava. The fog was so thick that a bulldozer operator could not see his own blade. Sigurdur Jonsson, who had lived all his life on Heimaey and at the time of the eruption had been working as a clerk in a hardware store, served from the outset with the pumping crews, and at times he was assigned to walk beside bulldozers and help guide them. Five people worked each dozer, he told me—the operator up in his cab, a person at either end of the blade, and the two others at the sides. The steel tracks of the big machines became so hot that they turned dark blue.

"One time, a bulldozer caught fire. Thirty meters away was a little pool where the water line had broken. He went into the pool. He came out and went to work again."

Sigurdur wore oversize boots with two or three pairs of wool socks.

"You couldn't stand for thirty seconds. In some areas, the lava was glowing. You could see a glow here, a glow there."

Not much below fifteen hundred degrees the lava stops glowing. Even if you can see it, you may not sense how hot it is.

"Leather boots shrank under the heat—Icelandic leather boots with lambskin, made for extreme cold. People carried water bottles to pour water on their shoes. We wore ski goggles, plastic goggles. Often, we could not see. You could not see an arm length. You found your way by the noise of the volcano, by the shape of the lava, by the pipelines. The pipelines guided us backward and forward. You used the pipeline all the way, working on the lava. Sometimes you saw the light of the sun through the fog. As the cooled area was getting bigger, visibility improved."

When the fog temporarily cleared, it was a straight shot of twenty miles across open water and up coastal slopes to the white walls of Eyjafjallajokull—at five thousand feet, a frozen cloud. A tenth of Iceland is capped with ice.

When Sigurdur worked at the harborside, spraying the advancing front, winter winds were fiercely cold.

"You were in sea spray. You got ice on your shoulders and chest. You went up on the lava and the ice would melt."

One day, seaweed plugged an intake for the pumps. He dived into the water and cleared away the seaweed.

The pumping platoon, at maximum, numbered seventy-five people, in a corps of five hundred who were engaged in the battle at any given time. Many came from Iceland (as the mainland is known on the island) to serve forty-eight hours at a stretch, and then go home. Sigurdur never left the island. He was, among other things, the head of the local chapter of Hjalparsveit skata—one of a number of first-aid-and-rescue groups, which are the closest thing to a militia known in Iceland—but he was not a leader in the fight against the lava. He was a foot soldier, and looked upon his role as such. Like the others, he viewed the struggle through the metaphor of war. "When the steam stops coming, and you see water running down the lava, you move the hose two or three meters. The lava front is black. Suddenly you get the red through the black. It could take hours—even days—to make it black. Sometimes it didn't work. You had to withdraw. You retreated, retreated until you were too close to the supply line; then you moved the supply line. In the battle, if you did not have to withdraw hoses and pipelines that was a victory for the day."

People behaved as if they were in combat, and on the streets of Reykjavik and other Icelandic towns veterans encountering one another will still talk about their service in Vestmannaeyjar.

The vapors of the eruption affected people's throats. Mouths were covered with cloth. "What the steam did to your throat—it made everybody hoarse," Sigurdur told me. "In the early days of the volcano, we were more hoarse. The doctors gave us medicine for it, which included Norwegian chest drops. You know what I mean by Norwegian chest drops."

"Ninety proof."

These recollections took place one summer evening in Sigurdur's house, Hasteinsvegi 47, in a close-set row on a tranquil street. Gentle and quiet in manner, a somewhat heavy man, partly bald now, with a mustache and gold-rimmed glasses, Sigurdur apologized frequently for his English, which he spoke at about the level of the average American professor. He had offered me a Heineken—a particularly generous gesture, but not so much of one that it would cause me to refuse. In Iceland, Norwegian chest drops—all the high hard ones from Jonny Walkersson to Jack Danielsson to Lambeater gin—are legal, and are sold over the counter at government stores. Beer, however, is prohibited, in the interest of the national health. The only legal beer brewed in Iceland is virtually nonalcoholic. In a somewhat paradoxical country, this is a savory paradox. Icelanders go to public bars that scrupulously obey the law. Not as a result of disrespect is theirs the oldest sitting democratic parliament in the world. Across the bar comes a bottle of prescription beer, accompanied routinely by a throat-burning shot of schnapps.

My Heineken in hand, I thanked Sigurdur warmly, and said, "It's hard to come by in Iceland."

He said, "It depends on what you do."

Continuing his narrative, he remarked that the eruption—for all its great surprise and early spectacle—had grown slowly.

"We had time to get used to it. You tried to protect your feet. I lost boots—I don't know how many. Only once I had to go to the doctor. Also your wrists. If you fell down, the wrist would be open between gloves and coat. So we wrapped our wrists with bandages. Very soon we stopped using helmets. Steam got inside helmets and hot water dripped on your head. We got from people's houses old gentlemen's hats with brims. They protected our faces. We had American Army helmets. We could always reach for them."

The helmets were for protection against falling ash and, in proximity to the crater, huge volcanic bombs. The ash was typically about as large as bits of pea gravel. They fell by the millions—hot enough and sharp enough to burn and cut skin. The bombs were ejections of molten lava that flew high into the air, became spherical, and, with contained gases, sometimes exploded like fireworks. In their interiors, bombs were generally molten. After they landed and broke, red-hot liquid poured out. Bombs that landed as much as two-thirds of a mile from the crater might weigh as much as sixty pounds. Bombs that carried half that distance could weigh a third of a ton. Sigurdur Jonsson and the others in the pumping crews usually worked outside the radius of the big bombs—except when they were on crater watch, in an ironclad hut quite close to the volcano, where forward observers could see into the churning lava, report its current style of eruption, and warn their colleagues in the fog about fresh eruptive flows.

I asked him, "What if a big fiery rock came down when you were working blind like that on the lava in the dense fog?"

He said, "It just didn't." ▪

Pliny the Younger

Account of the Death of Pliny the Elder (A.D. 24–79) written
by his nephew at the request of the historian Tacitus
translated from the Latin by Cynthia Damon

My dear Tacitus,

You ask me to write you something about the death of my uncle so that
the account you transmit to posterity is as reliable as possible. I am grateful to you, for I see that his death will be remembered forever if you write
about it [in your Histories]. He perished in a devastation of the loveliest of
lands, in a memorable disaster shared by peoples and cities, though this
will be a kind of eternal life for him. Although he himself wrote a great
number of enduring works, the imperishable nature of your writings will
add a great deal to the survival of his memory. Happy are they, in my opinion, to whom it is given either to do something worth writing about, or to
write something worth reading; most happy, of course, those who do both.
With his own books and yours, my uncle will be counted among the latter. It is therefore with great pleasure that I take up, or rather take upon
myself, the task you have set me.

He was at Misenum in his capacity as commander of the fleet on the
24th of August [A.D. 79], when, between two and three in the afternoon,
my mother drew his attention to a cloud of unusual size and appearance.
He had had a sunbath, then a cold bath, and was reclining after dinner with
his books. He called for his shoes and climbed up to where he could get the
best view of the phenomenon. The cloud was rising from a mountain—at
such a distance we couldn't tell which, but afterward learned that it was
Vesuvius. I can best describe its shape by likening it to a pine tree. It rose
into the sky on a very long "trunk" from which spread some "branches."
I imagine it had been raised by a sudden blast, which then weakened,
leaving the cloud unsupported so that its own weight caused it to spread
sideways. Some of the cloud was white; in other parts there were dark
patches of dirt and ash. The sight of it made the scientist in my uncle
determined to see it from closer at hand.

He ordered a boat made ready. He offered me the opportunity of going along, but I preferred to study—he himself happened to have set me a writing exercise. As he was leaving the house he was brought a letter from Tascius' wife, Rectina, who was terrified by the looming danger. Her villa lay at the foot of Vesuvius, and there was no way out except by boat. She begged him to get her away. He changed his plans. The expedition that started out as a quest for knowledge now called for courage. He launched the quadriremes and embarked himself, a source of aid for more people than just Rectina, for that delightful shore was a populous one. He hurried to a place from which others were fleeing and held his course directly into danger. Was he afraid? It seems not, as he kept up a continuous observation of the various movements and shapes of that evil cloud, dictating what he saw.

Ash was falling onto the ships now, darker and denser the closer they went. Now it was bits of pumice, and rocks that were blackened and burned and shattered by the fire. Now the sea is shoal; debris from the mountain blocks the shore. He paused for a moment wondering whether to turn back as the helmsman urged him. "Fortune helps the brave," he said. "Head for Pomponianus."

At Stabiae, on the other side of the bay formed by the gradually curving shore, Pomponianus had loaded up his ships even before the danger arrived, though it was visible and indeed extremely close once it intensified. He planned to put out as soon as the contrary wind let up. That very wind carried my uncle right in, and he embraced the frightened man and gave him comfort and courage. In order to lessen the other's fear by showing his own unconcern, he asked to be taken to the baths. He bathed and dined, carefree or at least appearing so (which is equally impressive). Meanwhile, broad sheets of flame were lighting up many parts of Vesuvius; their light and brightness were the more vivid for the darkness of the night. To alleviate people's fears my uncle claimed that the flames came from the deserted homes of farmers who had left in a panic with the hearth fires still alight. Then he rested and gave every indication of actually sleeping; people who passed by his door heard his snores, which were rather resonant since he was a heavy man. The ground outside his room rose so high with the mixture of ash and stones that if he had spent any more time there escape would have been impossible. He got up and came out, restoring himself to Pomponianus and the others who had been unable to sleep. They discussed what to do, whether to remain under cover or to try the open air. The buildings were being rocked by a series of strong tremors and appeared to have come loose from their foundations and to be sliding this way and that. Outside, however, there was danger from the rocks that were coming down, light and fire-consumed as these bits of pumice were. Weighing the relative dangers, they chose the outdoors; in my uncle's case it was a rational decision; others just chose the alternative that frightened them the least.

They tied pillows on top of their heads as protection against the shower of rock. It was daylight now elsewhere in the world, but there the darkness was darker and thicker than any night. But they had torches and other lights. They decided to go down to the shore, to see from close up if anything was possible by sea. But it remained as rough and uncooperative as before. Resting in the shade of a sail he drank once or twice from the cold water he had asked for. Then came a smell of sulfur, announcing the flames, and the flames themselves, sending others into flight but reviving him. Supported by two small slaves he stood up, and immediately collapsed. As I understand it, his breathing was obstructed by the dust-laden air, and his innards, which were never strong and often blocked or upset, simply shut down. When daylight came again two days after he died, his body was found untouched, unharmed, in the clothing that he had had on. He looked more asleep than dead.

Meanwhile at Misenum, my mother and I . . . but this has nothing to do with history, and you only asked for information about his death. I'll stop here then. But I will say one more thing, namely, that I have written out everything that I did at the time and heard while memories were still fresh. You will use the important bits, for it is one thing to write a letter, another to write history, one thing to write to a friend, another to write for the public. Farewell. ■

Charles Darwin

FROM *Voyage of the Beagle* (1845)

CHAPTER XIV.   *Chiloe and Concepcion: Great Earthquake*

On January the 15th [1835] we sailed from Low's Harbour, and three days afterwards anchored a second time in the bay of S. Carlos in Chiloe. On the night of the 19th the volcano of Osorno was in action. At midnight the sentry observed something like a large star, which gradually increased in size till about three o'clock, when it presented a very magnificent spectacle. By the aid of a glass, dark objects, in constant succession, were seen, in the midst of a great glare of red light, to be thrown up and to fall down. The light was sufficient to cast on the water a long bright reflection. Large masses of molten matter seem very commonly to be cast out of the craters in this part of the Cordillera. I was assured that when the Corcovado is in eruption, great masses are projected upwards and are seen to burst in the air, assuming many fantastical forms, such as trees: their size must be immense, for they can be distinguished from the high land behind S. Carlos, which is no less than ninety-three miles from the Corcovado. In the morning the volcano became tranquil.

I was surprised at hearing afterwards that Aconcagua in Chile, 480 miles northwards, was in action on this same night; and still more surprised to hear, that the great eruption of Coseguina (2700 miles north of Aconcagua), accompanied by an earthquake felt over 1000 miles, also occurred within six hours of this same time. This coincidence is the more remarkable, as Coseguina had been dormant for twenty-six years: and Aconcagua most rarely shows any signs of action. It is difficult even to conjecture, whether this coincidence was accidental, or shows some subterranean connexion. If Vesuvius, Etna, and Hecla in Iceland (all three relatively nearer each other, than the corresponding points in South America) suddenly burst forth in eruption on the same night, the coincidence would be thought remarkable; but it is far more remarkable in this case, where the three vents fall on the same great mountain-chain, and where the vast plains along the entire eastern coast, and the upraised recent shells along more than 2000 miles on the western coast, show in how equable and connected a manner the elevatory forces have acted.

CHAPTER XVII. *Galapagos Archipelago*

September 15th.—This archipelago consists of ten principal islands, of which five exceed the others in size. They are situated under the Equator, and between five and six hundred miles westward of the coast of America. They are all formed of volcanic rocks; a few fragments of granite curiously glazed and altered by the heat, can hardly be considered as an exception. Some of the craters, surmounting the larger islands, are of immense size, and they rise to a height of between three and four thousand feet. Their flanks are studded by innumerable smaller orifices. I scarcely hesitate to affirm, that there must be in the whole archipelago at least two thousand craters. These consist of either lava or scoriae, or of finely-stratified, sandstone-like tuff. [. . .]

Considering that these islands are placed directly under the equator, the climate is far from being excessively hot; this seems chiefly caused by the singularly low temperature of the surrounding water, brought here by the great southern Polar current. Excepting during one short season very little rain falls and even then it is irregular; but the clouds generally hang low. Hence, whilst the lower parts of the island are very sterile, the upper parts, at a height of a thousand feet and upwards, possess a damp climate and a tolerably luxuriant vegetation. This is especially the case on the windward sides of the islands, which first receive and condense the moisture from the atmosphere.

In the morning (17th) we landed on Chatham Island, which, like the others, rises with a tame and rounded outline, broken here and there by scattered hillocks, the remains of former craters. Nothing could be less inviting than the first appearance. A broken field of black basaltic lava, thrown into the most rugged waves, and crossed by great fissures, is every where covered by stunted, sunburnt brushwood, which shows little signs of life. The dry and parched surface, being heated by the noonday sun, gave to the air a close and sultry feeling, like that from a stove: we fancied even that the bushes smelt unpleasantly. Although I diligently tried to collect as many plants as possible, I succeeded in getting very few; and such wretched-looking little weeds would have better become an arctic than an equatorial Flora. [. . .]

The *Beagle* sailed round Chatham Island, and anchored in several bays. One night I slept on shore on a part of the island, where black truncated cones were extraordinarily numerous: from one small eminence I counted sixty of them, all surmounted by craters more or less perfect. The greater number consisted merely of a ring of red scoriae or slags, cemented together: and their height above the plain of lava was not more than from fifty to a hundred feet; none had been very lately active. The entire surface of this part of the island seems to have been permeated, like a sieve, by the subterranean vapours; here and there the lava, whilst soft, has been blown into great bubbles; and in other parts, the tops of caverns similarly formed have fallen in, leaving circular pits with steep sides. From the regular form of the many craters, they gave to the country an artificial appearance, which vividly reminded me of those parts of Staffordshire, where the great iron-foundries are most numerous. The day was glowing hot, and the scrambling over the rough surface and through the intricate thickets, was very fatiguing; but I was well repaid by the strange Cyclopean scene. As I was walking along I met two large tortoises, each of which must have weighed at least two hundred pounds: one was eating a piece of cactus, and as I approached, it stared at me and slowly stalked away; the other gave a deep hiss, and drew in its head. These huge reptiles, surrounded by the black lava, the leafless shrubs, and large cacti, seemed to my fancy like some antediluvian animals. The few dull-coloured birds cared no more for me, than they did for the great tortoises.

[. . .]

The natural history of these islands is eminently curious, and well deserves attention. Most of the organic productions are aboriginal creations, found nowhere else; there is even a difference between the inhabitants of the different islands; yet all show a marked relationship with those of America, though separated from that continent by an open space of ocean, between 500 and 600 miles in width. The archipelago is a little world within itself, or rather a satellite attached to America, whence it has derived a few stray colonists, and has received the general character of its indigenous productions. Considering the small size of these islands, we feel the more astonished at the number of their aboriginal beings, and at their confined range. Seeing every height crowned with its crater, and the boundaries of most of the lava-streams still distinct, we are led to believe that within a period, geologically recent, the unbroken ocean was here spread out. Hence, both in space and time, we seem to be brought somewhat near to that great fact—that mystery of mysteries—the first appearance of new beings on this earth. ▪

Annie Dillard

FROM "Life on the Rocks: The Galápagos"
  in *Teaching a Stone to Talk* (1982)

First there was nothing, and although you know with your reason that
nothing is nothing, it is easier to visualize it as a limitless slosh of sea—
say, the Pacific. Then energy contracted into matter, and although you
know that even an invisible gas is matter, it is easier to visualize it as a
massive squeeze of volcanic lava spattered inchoate from the secret pit of
the ocean and hardening mute and intractable on nothing's lapping shore—
like a series of islands, an archipelago. Like: the Galápagos. Then a softer
strain of matter began to twitch. It was a kind of shaped water; it flowed,
hardening here and there at its tips. There were blue-green algae; there
were tortoises.

The ice rolled up, the ice rolled back, and I knelt on a plain of lava boulders in the islands called Galápagos, stroking a giant tortoise's neck. The
tortoise closed its eyes and stretched its neck to its greatest height and vul-
nerability. I rubbed that neck and when I pulled away my hand, my palm
was green with a slick of single-celled algae. I stared at the algae, and at the
tortoise, the way you stare at any life on a lava flow, and thought: Well—
here we all are.

Being here is being here on the rocks. These Galapagonian rocks, one of
them seventy-five miles long, have dried under the equatorial sun between
five and six hundred miles west of the South American continent; they lie
at the latitude of the Republic of Ecuador, to which they belong.

There is a way a small island rises from the ocean affronting all reason.
It is a chunk of chaos pounded into visibility *ex nihilo:* here rough, here
smooth, shaped just so by a matrix of physical necessities too weird to contemplate, here instead of there, here instead of not at all. It is a fantastic
utterance, as though I were to open my mouth and emit a French horn, or
a vase, or a knob of tellurium. It smacks of folly, of first causes.

I think of the island called Daphnecita, little Daphne, on which I never set foot. It's in half of my few photographs, though, because it obsessed me; a dome of gray lava like a pitted loaf, the size of the Plaza Hotel, glazed with guano and crawling with red-orange crabs. Sometimes I attributed to this island's cliff face a surly, infantile consciousness, as though it were sulking in the silent moment after it had just shouted, to the sea and the sky, "I didn't ask to be born." Or sometimes it aged to a raging adolescent, a kid who's just learned that the game is fixed, demanding, "What did you have me for, if you're just going to push me around?" Daphnecita: again, a wise old island, mute, leading the life of pure creaturehood open to any antelope or saint. After you've blown the ocean sky-high, what's there to say? What if we the people had the sense or grace to live as cooled islands in an archipelago live, with dignity, passion, and no comment? ■

Haroun Tazieff

"Not a Very Sensible Place for a Stroll" in *Craters of Fire* (1952)
translated from the French by Eithne Wilkins

> I tossed off an immense gulp of poison.
> —*Rimbaud*, Une saison en Enfer

Standing on the summit of the growling cone, even before I got my breath back after the stiff climb, I peered down into the crater.

I was astonished. Two days previously the red lava had been boiling up to the level of the gigantic lip; now the funnel seemed to be empty. All that incandescent magma had disappeared, drawn back into the depths by the reflux of some mysterious ebb and flow, a sort of breathing. But there, about fifty feet below where I was standing, was the glow and the almost animate fury of the great throat which volcanologists call the conduit or chimney. It was quite a while before I could tear my eyes away from that lurid, fiery center, that weird palpitation of the abyss. At intervals of about a minute, heralded each time by a dry clacking, bursts of projectiles were flung up, running away up into the air, spreading out fan-wise, all aglare, and then falling back, whistling, on the outer sides of the cone. I was rather tense, ready to leap aside at any moment, as I watched these showers, with their menacing trajectories.

Each outburst of rage was followed by a short lull. Then heavy rolls of brown and bluish fumes came puffing out, while a muffled grumbling, rather like that of some monstrous watch-dog, set the whole bulk of the volcano quivering. There was not much chance for one's nerves to relax, so swiftly did each follow on the other—the sudden tremor, the burst, the momentary intensification of the incandescence, and the outbreak of a fresh salvo. The bombs went roaring up, the cone of fire opening out overhead, while I hung in suspense. Then came the hissing and sizzling, increasing in speed and intensity, each "whoosh" ending up in a muffled thud as the bomb fell. On their black bed of scoriae, the clots of molten magma lay with the fire slowly dying out of them, one after the other growing dark and cold.

Some minutes of observation were all I needed. I noted that today, apart from three narrow zones to the west, north, and north-east, the edges of the crater had scarcely been damaged at all by the barrage from underground. The southern point where I stood was a mound rising some twelve or fifteen feet above the general level of the rim, that narrow, crumbling lip of scoriae nearer to the fire, where I had never risked setting foot. I looked at this rather alarming ledge all round the crater, and gradually felt an increasing desire to do something about it . . . It became irresistible. After all, as the level of the column of lava had dropped to such an exceptional degree, was this not the moment to try what I was so tempted to do and go right round the crater?

Still, I hesitated. This great maw, these jaws sending out heat that was like the heavy breathing of some living creature, thoroughly frightened me. Leaning forward over that hideous glow, I was no longer a geologist in search of information, but a terrified savage.

"If I lose my grip," I said aloud, "I shall simply run for it."

The sound of my own voice restored me to normal awareness of myself. I got back my critical sense and began to think about what I could reasonably risk trying. "De l'audace, encore de l'audace. . . ." That was all very well, of course, but one must also be careful. Past experience whispered a warning not to rush into anything blindly. Getting the upper hand of both anxiety and impatience, I spent several minutes considering, with the greatest of care, the monster's manner of behaving. Solitude has got me into the habit of talking to myself, and so it was more or less aloud that I gave myself permission to go ahead.

"Right, then. It can be done."

I turned up my collar and buttoned my canvas jacket tight at the throat—I didn't want a sly cinder down the back of my neck! Then I tucked what was left of my hair under an old felt hat that did service for a helmet. And now for it!

Very cautiously indeed, I approach the few yards of pretty steep slope separating the peak from the rim I am going to explore. I cross, in a gingerly manner, a first incandescent crevasse. It is intense orange in color and quivering with heat, as though opening straight into a mass of glowing embers. The fraction of a second it takes me to cross it is just long enough for it to scorch the thick cord of my breeches. I get a strong whiff of burnt wool.

A promising start, I must say!

Here comes a second break in the ground. Damn it, a wide one, too! I can't just stride across this one: I'll have to jump it. The incline makes me thoughtful. Standing there, I consider the unstable slope of scoriae that will have to serve me for a landing-ground. If I don't manage to pull up . . . if I go rolling along down this funnel with the flames lurking at the bottom of it . . . My little expedition all at once strikes me as thoroughly rash, and I stay where I am, hesitating. But the heat under my feet is becoming unbearable. I can't endure it except by shifting around. It only needs ten seconds of standing still on this enemy territory, with the burning gases slowly and steadily seeping through it, and the soles of my feet are already baking hot. From second to second the alternative becomes increasingly urgent: I must jump for it or retreat.

Here I am! I have landed some way down the fissure. The ashes slide underfoot, but I stop without too much trouble. As so often happens, the anxiety caused by the obstacle made me over-estimate its importance.

Step by step, I set out on my way along the wide wall of slag-like debris that forms a sort of fortification all round the precipice. The explosions are still going on at regular intervals of between sixty and eighty seconds. So far no projectile had come down on this side, and this cheers me up considerably. With marked satisfaction I note that it is pretty rare for two bombs of the same salvo to fall less than three yards apart: the average distance between them seems to be one of several paces. This is encouraging. One of the great advantages of this sort of bombardment, compared with one by artillery, lies in the relative slowness with which the projectiles fall, the eye being able to follow them quite easily. Furthermore, these shells don't burst. But what an uproar, what an enormous, prolonged bellowing accompanies their being hurled out of the bowels of the earth!

I make use of a brief respite in order to get quickly across the ticklish north-eastern sector. Then I stop for a few seconds, just long enough to see yet another burst gush up and come showering down a little ahead of me, after which I start out for the conquest of the northern sector. Here the crest narrows down so much that it becomes a mere ridge, where walking is so difficult and balancing so precarious that I find myself forced to go on along the outer slope, very slightly lower down. Little by little, as I advance through all this tumult, a feeling of enthusiasm is overtaking me. The immediate imperative necessity for action has driven panic far into the background. And under the hot, dry skin of my face, taut on forehead and cheekbones, I can feel my compressed lips parting, of their own accord, in a smile of sheer delight. But look out!

A sudden intensification of the light warns me that I am approaching a point right in the prolongation of the fiery chimney. In fact, the chimney is not vertical, but slightly inclined in a north-westerly direction, and from here one can look straight down into it. These tellurian entrails, brilliantly yellow, seem to be surging with heat. The sight is so utterly amazing that I stand there, transfixed.

Suddenly, before I can make any move, the dazzling yellow changes to white, and in the same instant I feel a muffled tremor all through my body and there is a thunderous uproar in my ears. The burst of incandescent blocks is already in full swing. My throat tightens as, motionless, I follow with my gaze the clusters of red lumps rising in slow, perfect curves. There is an instant of uncertainty. And then down comes the hail of fire.

This time the warning was too short: I am right in the middle of it all. With my shoulders hunched up, head drawn back, chin in air, buttocks as much tucked in as possible, I peer up into the vault of sinister whining and whizzing there above me. All around bombs are crashing down, still pasty and soft, making a succession of muffled *plops*. One dark mass seems to have singled me out and is making straight for my face. Instinctively I take a leap to one side, and *feel* the great lump flatten itself out a few inches from my left foot. I should like to have a look, but this is not the moment! Here comes another projectile. I take another leap to dodge it. It lands close beside me. Then suddenly the humming in the air begins to thin out. There are a few more whizzing sounds, and then the downpour is over.

Have you ever tried to imagine a snail's state of mind as it creeps out of its shell again, the danger past? That was the way my head, which had been drawn back between hunched-up shoulders, gradually began to rise up again on my neck, and my arched back began to straighten, my arms to loosen, my hands to unclench. Right, then—it's better not to hang about in this sector! So I set out again. By this time I have got round three-quarters of the crater, and am in the gap between the northern and western zones, which are those that get the worst pounding. From here I can get back on to the ridge proper.

I am now almost directly over the roaring chasm, and my gaze goes straight down into it like a stone dropping into the pit. After all, it's nothing but a tunnel. That's all. It's a vertical tunnel, ten or fifteen yards across, its walls heated to such a degree that they stretch and "rise" like dough, and up from its depths every now and then enormous drops of liquid fire spurt forth, a great splashing sweat that falls and vanishes, golden flash upon flash, back into the dazzling gulf. Even the brownish vapors emanating from the pit cannot quite veil its splendor. It is nothing but a tunnel running down into viscous copper-colored draperies; yet it opens into the very substance of another world. The sight is so extraordinary that I forget the insecurity of my position and the hellish burning under the soles of my feet. Quite mechanically, I go on lifting first the left foot, then the right. It is as though my mind were held fast in a trap by the sight of this burning well from which a terrifying snore continually rises, interrupted by sharp explosions and the rolling of thunder.

Suddenly I hurl myself backwards. The flight of projectiles has whizzed past my face. Hunched up again, instinctively trying to make as small a target of myself as I can, I once more go through the horrors that I am beginning to know. I am in the thick of this hair's-breadth game of anticipation and dodging.

And now it's all over; I take a last glance into the marvellous and terrible abyss, and am just getting ready to start off on the last stage of this burning circumnavigation, all two hundred yards of it, when I get a sudden sharp blow in the back. A delayed-action bomb! With all the breath knocked out of me, I stand rigid.

A moment passes. I wonder why I am not dead. But nothing seems to have happened to me—no pain, no change of any sort. Slowly I risk turning my head, and at my feet I see a sort of huge red loaf with the glow dying out of it.

I stretch my arms and wriggle my back. Nothing hurts. Everything seems to be in its proper place. Later on, examining my jacket, I discovered a brownish scorch-mark with slightly charred edges, about the size of my hand, and I drew from it a conclusion of immense value to me in future explorations: so long as one is not straight in the line of fire, volcanic bombs, which fall in a still pasty state, but already covered with a kind of very thin elastic skin, graze one without having time to cause a deep burn.

I set off at a run, as lightly as my 165 pounds allow, for I must be as quick as I can in crossing this part of the crater-edge, which is one of the most heavily bombarded. But I am assailed by an unexpected blast of suffocating fumes. My eyes close, full of smarting tears. I am caught in a cloud of gas forced down by the wind. I fight for breath. It feels as if I were swallowing lumps of dry, corrosive cotton-wool. My head swims, but I urge myself at all costs to get the upper hand. The main thing is not to breathe this poisoned air. Groping, I fumble in a pocket. Damn, not this one. How about this other one, then? No. At last I get a handkerchief out and, still with my eyes shut, cover my mouth with it. Then, stumbling along, I try to get through the loathsome cloud. I no longer even bother to pay any attention to the series of bursts from the volcano, being too anxious to get out of this hell before I lose grip entirely. I am getting pretty exhausted, staggering . . . The air filtered through the handkerchief just about keeps me going, but it is still too poisonous, and there is too little of it for the effort involved in making this agonizing journey across rough and dangerous terrain. The gases are too concentrated, and the great maw that is belching them forth is too near.

A few steps ahead of me I catch a glimpse of the steep wall of the peak, or promontory, from the other side of which I started about a century ago, it seems to me now. The noxious mists are licking round the peak, which is almost vertical and twice the height of a man. It's so near! But I realize at once that I shall never have the strength to clamber up it.

In less than a second, the few possible solutions to this life-and-death problem race through my mind. Shall I turn my back to the crater and rush away down the outer slope, which is bombarded by the thickest barrages? No. About face and back along the ledge? Whatever I do, I must turn back. And then make my escape. By sliding down the northern slope? That is also under too heavy bombardment. And the worst of it would be that in making a descent of that sort there would be no time to keep a watch for blocks of lava coming down on one.

Only one possibility is left: to make my way back all along the circular ridge, more than a hundred yards of it, till I reach the eastern rim, where neither gas nor projectiles are so concentrated as to be necessarily fatal.

I swing round. I stumble and collapse on all fours, uncovering my mouth for an instant. The gulp of gas that I swallow hurts my lungs, and leaves me gasping. Red-hot scoriae are embedded in the palms of my hands. I shall never get out of this!

The first fifteen or twenty steps of this journey back through the acrid fumes of sulphur and chlorine are a slow nightmare; no step means any progress and no breath brings any oxygen into the lungs. The threat of bombs no longer counts. Only these gases exist now. Air! Air!

I came to myself again on the eastern rim, gasping down the clean air borne by the wind, washing out my lungs with deep fresh gulps of it, as though I could never get enough. How wide and comfortable this ledge is! What a paradise compared with the suffocating, torrid hell from which I have at last escaped! And yet this is where I was so anxious and so tense less than a quarter of an hour ago.

Several draughts of the prevailing breeze have relieved my agony. All at once, life is again worth living! I no longer feel that desire to escape from here as swiftly as possible. On the contrary, I feel a new upsurge of explorer's curiosity. Once more my gaze turns towards the mouth, out of which sporadic bursts of grape-shot are still spurting forth. Now and then there are bigger explosions and I have to keep a look-out for what may come down on my head, which momentarily interrupts the dance I keep up from one foot to the other, that *tresca* of which Dante speaks—the dance of the damned, harried by fire. True, I have come to the conclusion that the impact of these bombs is not necessarily fatal, but I am in no hurry to verify the observation.

The inner walls of the crater do not all incline at the same angle. To the north, west and south, they are practically vertical, not to say overhanging, but here on the east the slope drops away at an angle of no more than fifty degrees. So long as one moved along in a gingerly way, this might be an incline one could negotiate. It would mean going down into the very heart of the volcano. For an instant I am astounded by my own foolhardiness. Still, it's really too tempting. . . .

Cautiously, I take a step forward . . . then another . . . and another . . . seems all right . . . it *is* all right. I begin the climb down, digging my heels as deep as I can into the red-hot scoriae. Gradually below me, the oval of the enormous maw comes nearer, growing bigger, and the terrifying uproar becomes more deafening. My eyes, open as wide as they will go, are drunken with its monstrous glory. Here are those ponderous draperies of molten gold and copper, so near—so near that I feel as if I, human being that I am, had entered right into their fabulous world. The air is stifling hot. I am right in the fiery furnace.

I linger before this fascinating spectacle. But then, by sheer effort, I tear myself away. It's time to get back to being "scientific" and measure the temperatures, of the ground, and of the atmosphere. I plunge the long spike of the thermometer into the shifting scoriae, and the steel of it glitters among these brownish and grey screes with their dull shimmer. At a depth of six inches the temperature is two hundred and twenty degrees centigrade. It's amusing to think that when I used to dream it was always about polar exploration!

Suddenly, the monster vomits out another burst; so close that the noise deafens me. I bury my face in my arms. Fortunately almost every one of the projectiles comes down outside the crater. And now all at once I realize that it is I who am here—*alive* in this crater, surrounded by scorching walls, face to face with the very mouth of the fire. Why have I got myself into this trap, alone and without the slightest chance of help? Nobody in the world has any suspicion of the strange adventure on which I have embarked, and nobody, for that matter, could now do the slightest thing about it. Better not think about it . . .

Without a break the grim, steady growling continues to rise from the depths of that throat, only out-roared at intervals by the bellowing and belching of lava. It's too much; I can feel myself giving up. I turn my back on it, and try, on all fours, to scramble up the slope, which has now become incredibly steep and crumbles and gives way under my weight, which is dragging me down, down . . . "Steady, now," I say to myself. "Keep calm for a moment. Let's work it out. Let's work it out properly. Or else, my boy, this is the end of *you*."

Little by little, by immense exertions, I regain control of my movements, as well as the mental steadiness I need. I persuade myself to climb *calmly* up this short slope, which keeps crumbling away under my feet. When I reach the top, I stand upright for just a moment. Then, crossing the two glowing fissures that still intersect my course, I reach the part of the rim from where there is a way down into the world of ordinary peaceful things. ▪

James D. Houston

"Fire in the Night" (1990)

Among geologists and volcano buffs there is a little rite of passage, whereby you stick your hand axe into moving lava and bring away a gob of the molten stuff. In order to do this you have to be where the lava is flowing and hot, then you have to get your body in close enough to the heat to reach down toward the edge of the flow, and it usually means you have to walk or stand for at least a few seconds on some pretty thin crust.

My chance came one night last year, on the Big Island of Hawaii, when I hiked out along the southern shoreline toward the spilling end of a lava tube. I was traveling in the company of Jack Lockwood, a specialist in volcanic hazards with the U.S. Geological Survey. He is a trim and wiry fellow, with wild hair and a devilish grin, a man from New England who has found that island, its craters and its lava fields, to be his natural habitat. He loves it there, he loves the look of the ropey *pahoehoe*, the many shapes it takes. He will stop the car to study the way today's flow has poured over yesterday's, making a drapery of knobs and drips. He will remark upon the metallic sheen in the late sun, and then point out that newer lava can be crumbled with your shoe, while the stuff that came through yesterday has already hardened under a rainfall and thus is firmer.

We parked where the yellow line of the coast road disappeared under a ten-foot wall of new rock. We got out the packs, the gloves, the canteens, the flashlights, the hard-hats. Jack's hat was custom-made, with his name in raised letters on the metal. His hard-toe boots were scuffed ragged with threads of rock-torn leather. I was going to wear running shoes for this expedition, until he told me no. "Where we're going," he said, "the soles could peel right off."

Hunkered on the asphalt, lacing up the high-top boots I'd borrowed, I could already feel it shimmering toward us. Minutes later we were hiking through furnace heat, over lava that had rolled across there just a few hours earlier. Through cracks and fissures you could see the molten underlayer showing, three or four inches below the dark surface.

"You can actually walk on it fifteen or twenty minutes after it starts to harden," Jack said, "as long as you have an inch of surface underfoot."

Soon the red slits were everywhere, and we were crossing what appeared to be several acres of recent flow. Jack plunged ahead with great purpose, with long firm strides, planting each foot and leaning forward as he walked, as if there were a path to be followed and we were on it—though of course there was no path, no prior footprints, no markers of any kind to guide us across terrain that had not been there that morning.

"Jack," I said, "have you ever stepped into a soft spot? I mean, got burned, fallen through?"

He shook his head vigorously. "Nope."

"How do you know where to step?"

He stopped and looked at me with his mischievous eyes, his beard and his squint reminding me of a young John Huston. "You just pick your way and pay attention as you go. It's partly experience and partly faith."

"Faith?"

"You have to put your trust in Pele. Tell her that you come out here with respect, and she will take care of you."

As he plunged on, I wanted to trust in Pele, the goddess of fire, who is said to make her home in a crater about fifteen miles from where we were walking. We had already talked about her, while driving down Chain of Craters Road, and I knew he meant what he'd just said. But I have to confess that at the moment I was putting my full trust in Lockwood, placing my feet where he placed his, stepping in his steps as we strode and leaped from rock to rock.

Eventually the heat subsided, and we were hiking over cooler terrain, though none of it was very old. "Everything you see has flowed through in the last six months," he said. Two and a half miles of the coast road had recently been covered, as well as the old settlement of Kamoamoa, near where we'd parked. Inland we could see some of what remained of Royal Gardens, a subdivision laid out in the early 1970s, laid out right across a slope of the East Rift Zone. In the Royal Gardens grid, cross-streets had been named for tropical flowers—Gardenia, Pikake—while the broader main streets sounded noble—Kamehameha, Prince. Now the access road was blocked in both directions. From our vantage point it looked as if great vats of black paint had been dumped over the highest ridge, to pour down the slope and through the trees, to cover boulevards and lawns.

Our destination—the spilling tube—was marked by a steam plume rising high against the evening sky. When we left the car it was white and feathery at the top, two miles down the coast. After the sun set and the light began to dim, the plume turned pink and red. Spatter thrown up from the collision of lava and surf had formed a littoral cone now outlined against the steam. As we approached, tiny figures could be seen standing at the edge of this cone, like cut-outs against the fiery backdrop.

On one side of the cone, flat spreads of lava were oozing toward the cliff. On the other side, an orange gusher was arcing some thirty feet above the water, while a mound slowly rose beneath it. Beyond that tube, another spill obscured by steam sent lava straight into the water at about sea level. Red-and-black floating gobs spewed out from the steam, or sometimes flew into the air, breaking into fiery debris that was gradually building the littoral cone.

These fires lit the billowing plume from below. As it churned away toward the west, it sent a pinkish glow back down onto the marbled stuff, which made me think of the Royal Hawaiian Hotel, where they spend a lot of money on light bulbs and filters trying to tint the offshore waters a Waikiki pink that can never come close to Pele's cosmetic kit.

A video cameraman was there, perched at the cliff filming the build-up on the mound below the arching orange tube. His tripod legs were spindly black against the glow. Nearby a couple of dozen people stood gazing at the spectacle, staffers from Volcano Observatory and the University of Hawaii. They were out there in numbers, Jack told me later, because this was a rare night. Spills like this were usually closer to the surf, and the lava would pour until the mound built up from below to seal off its opening. But this littoral cone was unstable, and part of it had fallen away and beheaded the end of the tube, so the lava was spilling free from high up the cliff, making a liquid column of endlessly sizzling orange.

If you could take your eyes off its mesmerizing arc and turn inland, you could see another glow in the night. It hung above the nearest ridge, light from the lake called Kupaianaha, the source of the lava moving around us. It was a new lake, inside a new shield cone. From there the lava snaked seaward via a channel that looped wide to the east, then back toward where we stood. You could see evidence of its twisting, subterranean path about halfway down the mountain, where tiny fires seemed to be burning, four or five eyes of flame against the black.

We lingered for an hour, maybe more, chatting, bearing witness, sharing our wonder with the others lucky enough to be out there on such a night, at the cutting edge of destruction and creation. We were about to start back when Lockwood said this is probably as good a time as any for me to add my name to the "one thousandth of one percent of the human population who had stuck their axe in hot lava." And with that he began to prowl around a couple of oozing streams, to see how close we could get.

I watched him step out onto some hot stuff that had barely stopped moving, and saw the surface give under his boot. With a grin he jumped back. "That's probably a little too soft."

We moved around to the far side, forty feet away, and approached the fiery mush from another angle. With axe in hand he hopped across the one-inch crust and dug into the front edge of a narrow strip, but it was already cooling and a little too thick to lift. He could only pull it up an inch or so, the front lip already in that halfway zone between liquid and solid stone.

He was pulling so hard he lost his footing and half fell toward the crust. His gloved hand reached out to take the fall, and for a moment his crouching body was silhouetted against the molten stream, while behind him the red-and-orange steam plume surged like a backdrop curtain for his dance. He came rolling and hopping toward me with a wild grin and a rascal eye.

"That's a little too viscous. It's surprising. It's cooler than it looks."

So we moved on, heading back the way we'd come, under a black sky with its infinity of stars, our flashlight beams bobbing across the rocks, while the plume grew smaller behind us.

We dropped down to a new beach of dark volcanic sand, then climbed out of the sand onto that day's fresh lava, where the red slits once again glowed all around us. As we picked our way, in the furnace heat, we came upon a flow that had not been there when we crossed the first time.

"Pele is being good to you," said Jack, grinning, his beard red-tinted underneath. He handed me his axe. "This is perfect. Just keep your back to the heat, and move in quickly."

Which is what I did. The stream was maybe twenty feet wide, seething, creeping toward the sea. I back pedaled up next to it, reaching with the flat chisel end of the metal blade, dipped and scooped into the burning lip. It was smoother than wet cement, thicker than honey, thicker than three-finger poi. Maybe the consistency of glazing compound, or the wet clay potters use. For the first mini-second it felt that way. As I dug in and pulled, it was already harder. It clung to the flow, but I tugged and finally came away with a chunk the size of a tennis ball, which held to the blade as I leaped back away from the heat.

Jack was excited. "Throw it down here, quickly!"

I plopped it between us, on a black slab.

"Now press your heel in hard!"

I pressed my boot heel into the glob, flattening it with a boot print. When the rubber began to smoke, I pulled my foot away.

"Now," he said, with a happy grin, "we'll put this on my shovel blade and carry it to the car while it cools, and this will be your souvenir."

By the time we reached the asphalt road the heat had given way to balmy coastal air off the water. The slits and fissures and plumes and flows were all behind us, and that was the end of our expedition.

But it was not the end of my relationship with that flattened piece of rock. I lived with it for another week, trying to decide what to do. After such a magical night, the idea of a souvenir appealed to me. It was mine, I suppose, because I had marked it with my boot. Now it was smooth, as shiny as black glass, and if I lived on that island I'd probably have it sitting on my desk. But I did not feel right about bringing this trophy back home. I kept thinking about the tug of the lava as I pulled the axe away. Through the handle I had felt its texture, its consistency, and something else that haunted me. A reluctance. A protest. As if live flesh were being torn from a body.

Maybe this was what the Hawaiians meant when they said all the rocks there belonged to Pele and should not leave the island. Maybe the unwritten law that said be respectful of the rocks was another way of honoring that old yearning in the stone. Maybe Pele was another word for the living stuff of earth, and maybe I had finally understood something, through my hands, something I had heard about and read about and talked about and even tried to write about.

Before I left the island I drove down to the south shore again. Sighting from the new black sand beach, I think I got pretty close to where we'd been. I dropped the chunk of lava down into a jagged crevice and asked it to forgive me for any liberties I might have taken, and I thanked Pele for letting me carry this rock around for a while. Then I drove to the airport and checked in my rental car and caught my plane.

I don't tell people about this, by the way. Not here on the mainland. You come back to California and tell someone you have been talking to rocks, they give you a certain kind of look. I'll mention it to Jack Lockwood, of course, the next time I see him. It's easier to talk about when you're in the islands. You meet a lot of people over there who claim to be on speaking terms with rocks. When you're in or near volcano country, it's easier to remember that they too have life, that each rock was once a moving thing, as red as blood and making eyes of fire in the night. ▪

Ursula K. Le Guin

"A Very Warm Mountain" (1980)

Everybody takes it personally. Some get mad. Damn stupid mountain went and dumped all that dirty gritty glassy gray ash that flies like flour and lies like cement all over their roofs, roads, and rhododendrons. Now they have to clean it up. And the scientists are a real big help, all they'll say is we don't know, we can't tell, she might dump another load of ash on you just when you've got it all cleaned up. It's an outrage.

Some take it ethically. She lay and watched her forests being cut and her elk being hunted and her lakes being fished and fouled and her ecology being tampered with and the smoky, snarling suburbs creeping closer to her skirts, until she saw it was time to teach the White Man's Children a lesson. And she did. In the process of the lesson, she blew her forests to matchsticks, fried her elk, boiled her fish, wrecked her ecosystem, and did very little damage to the cities: so that the lesson taught to the White Man's Children would seem, at best, equivocal.

But everybody takes it personally. We try to reduce it to human scale. To make a molehill out of the mountain.

Some got very anxious, especially during the dreary white weather that hung around the area after May 18 (the first great eruption, when she blew 1300 feet of her summit all over Washington, Idaho, and points east) and May 25 (the first considerable ashfall in the thickly populated Portland area west of the mountain). Farmers in Washington State who had the real fallout, six inches of ash smothering their crops, answered the reporters' questions with polite stoicism; but in town a lot of people were cross and dull and jumpy. Some erratic behavior, some really weird driving. "Everybody on my bus coming to work these days talks to everybody else, they never used to." "Everybody on my bus coming to work sits there like a stone instead of talking to each other like they used to." Some welcomed the mild sense of urgency and emergency as bringing people together in mutual support. Some—the old, the ill—were terrified beyond reassurance. Psychologists reported that psychotics had promptly incorporated the volcano into their private systems; some thought they were controlling her, and some thought she was controlling them. Businessmen, whom we know from the Dow Jones Reports to be an almost ethereally timid and emotional breed, read the scare stories in Eastern newspapers and cancelled all their conventions here; Portland hotels are having a long cool summer. A Chinese Cultural Attaché, evidently preferring earthquakes, wouldn't come farther north than San Francisco. But many natives were irrationally exhilarated, secretly, heartlessly welcoming every steam-blast and earth-tremor: Go it, mountain!

Everybody read in the newspaper everywhere that the May 18 eruption was "five hundred times greater than the bomb dropped on Hiroshima." Some reflected that we have bombs much more than five hundred times more powerful than the 1945 bombs. But these are never mentioned in the comparisons. Perhaps it would upset people in Moscow, Idaho or Missoula, Montana, who got a lot of volcanic ash dumped on them, and don't want to have to think, what if that stuff had been radioactive? It really isn't nice to talk about, is it. I mean, what if something went off in New Jersey, say, and *was* radioactive—Oh, stop it. That volcano's way out west there somewhere anyhow.

Everybody takes it personally.

I had to go into hospital for some surgery in April, while the mountain was in her early phase—she jumped and rumbled, like the Uncles in *A Child's Christmas in Wales*, but she hadn't done anything spectacular. I was hoping she wouldn't perform while I couldn't watch. She obliged and held off for a month. On May 18 I was home, lying around with the cats, with a ringside view: bedroom and study look straight north about forty-five miles to the mountain.

I kept the radio tuned to a good country western station and listened to the reports as they came in, and wrote down some of the things they said. For the first couple of hours there was a lot of confusion and contradiction, but no panic, then or later. Late in the morning a man who had been about twenty miles from the blast described it: "Pumice-balls and mud-balls began falling for about a quarter of an hour, then the stuff got smaller, and by nine it was completely and totally black dark. You couldn't see ten feet in front of you!" He spoke with energy and admiration. Falling mud-balls, what next? The main west coast artery, I-5, was soon closed because of the mud and wreckage rushing down the Toutle River towards the highway bridges. Walla Walla, 160 miles east, reported in to say their street lights had come on automatically at about ten in the morning. The Spokane-Seattle highway, far to the north, was closed, said an official expressionless voice, "on account of darkness."

At one-thirty that afternoon, I wrote:

It has been warm with a white high haze all morning, since six A.M., when I saw the top of the mountain floating dark against yellow-rose sunrise sky above the haze.

That was, of course, the last time I saw or will ever see that peak.

Now we can see the mountain from the base to near the summit. The mountain itself is whitish in the haze. All morning there has been this long, cobalt-bluish drift to the east from where the summit would be. And about ten o'clock there began to be visible clots, like cottage cheese curds, above the summit. Now the eruption cloud is visible from the summit of the mountain till obscured by a cloud layer at about twice the height of the mountain, i.e., 25–30,000 feet. The eruption cloud is very solid-looking, like sculptured marble, a beautiful blue in the deep relief of baroque curls, sworls, curled-cloud-shapes— darkening towards the top—a wonderful color. One is aware of motion, but (being shaky, and looking through shaky binoculars) I don't actually see the carven-blue-sworl-shapes move. Like the shadow on a sundial. It is enormous. Forty-five miles away. It is so much bigger than the mountain itself. It is silent, from this distance. Enormous, silent. It looks not like anything earthy, from the earth, but it does not look like anything atmospheric, a natural cloud, either. The blue of it is storm-cloud blue but the shapes are far more delicate, complex, and immense than stormcloud shapes, and it has this solid look; a weightiness, like the capital of some unimaginable column—which in a way indeed it is, the pillar of fire being underground.

At four in the afternoon a reporter said cautiously, "Earthquakes are being felt in the metropolitan area," to which I added, with feeling, "I'll say

they are!" I had decided not to panic unless the cats did. Animals are supposed to know about earthquakes, aren't they? I don't know what our cats
know; they lay asleep in various restful and decorative poses on the swaying floor and the jiggling bed, and paid no attention to anything except dinner time. I was not allowed to panic.

At four-thirty a meteorologist, explaining the height of that massive
storm-blue pillar of cloud, said charmingly, "You must understand that the
mountain is very warm. Warm enough to lift the air over it to 75,000
feet."

And a reporter: "Heavy mud flow on Shoestring Glacier, with continuous lightning." I tried to imagine that scene. I went to the television, and
there it was. The radio and television coverage, right through, was splendid. One forgets the joyful courage of reporters and cameramen when there
is something worth reporting, a real Watergate, a real volcano.

On the 19th, I wrote down from the radio, "A helicopter picked the log-
ger up while he was sitting on a log surrounded by a mud flow." This res-
cue was filmed and shown on television: the tiny figure crouching hope-
less in the huge abomination of ash and mud. I don't know if this man
was one of the loggers who later died in the Emanuel Hospital burn center, or if he survived. They were already beginning to talk about the
"killer eruption," as if the mountain had murdered with intent. Taking
it personally . . . Of course she killed. Or did they kill themselves? Old
Harry who wouldn't leave his lodge and his whiskey and his eighteen cats
at Spirit Lake, and quite right too, at eighty-three; and the young cameraman and the young geologist, both up there on the north side on the
job of their lives; and the loggers who went back to work because logging
was their living; and the tourists who thought a volcano is like Channel
Six, if you don't like the show you turn it off, and took their RVs and their
kids up past the roadblocks and the reasonable warnings and the weary
county sheriffs sick of arguing: they were all there to keep the appointment. Who made the appointment?

A firefighter pilot that day said to the radio interviewer, "We do what
the mountain says. It's not ready for us to go in."

On the 21st I wrote:

Last night a long, strange, glowing twilight; but no ash has yet fallen west of the mountain. Today, fine, gray, mild, dense Oregon rain. Yesterday afternoon we could see her vaguely through the glasses. Looking appallingly lessened—short, flat—That is painful. She was so beautiful. She hurled her beauty in dust clear to the Atlantic shore, she made sunsets and sunrises of it, she gave it to the western wind. I hope she erupts magma and begins to build herself again. But I guess she is still unbuilding. The Pres. of the U.S. came today to see her. I wonder if he thinks he is on her level. Of course he could destroy much more than she has destroyed if he took a mind to.

On June 4 I wrote:

Could see her through the glasses for the first time in two weeks or so. It's been dreary white weather with a couple of hours sun in the afternoons.—Not the new summit, yet; that's always in the roil of cloud/plume. But both her long lovely flanks. A good deal of new snow has fallen on her (while we had rain), and her SW face is white, black, and gray, much seamed, in unfamiliar patterns.

"As changeless as the hills—"

Part of the glory of it is being included in an event on the geologic scale. Being enlarged. "I shall lift up mine eyes unto the hills," yes: "whence cometh my help."

In all the Indian legends dug out by newspaper writers for the occasion, the mountain is female. Told in the Dick-and-Jane style considered appropriate for popular reportage of Indian myth, with all the syllables hyphenated, the stories seem even more naive and trivial than myths out of context generally do. But the theme of the mountain as woman—first ugly, then beautiful, but always a woman—is consistent. The mapmaking whites of course named the peak after a man, an Englishman who took his title, Baron St. Helens, from a town in the North Country: but the name is obstinately feminine. The Baron is forgotten, Helen remains. The whites who lived on and near the mountain called it The Lady. Called her The Lady. It seems impossible not to take her personally. In twenty years of living through a window from her I guess I have never really thought of her as "it."

She made weather, like all single peaks. She put on hats of cloud, and took them off again, and tried a different shape, and sent them all skimming off across the sky. She wore veils: around the neck, across the breast: white, silver, silver-gray, gray-blue. Her taste was impeccable. She knew the weathers that became her, and how to wear the snow.

Dr. William Hamilton of Portland State University wrote a lovely piece for the college paper about "volcano anxiety," suggesting that the silver cone of St. Helens had been in human eyes a breast, and saying:

> St. Helens' real damage to us is not . . . that we have witnessed a denial of the trustworthiness of God (such denials are our familiar friends). It is the perfection of the mother that has been spoiled, for part of her breast has been removed. Our metaphor has had a mastectomy.
>
> At some deep level, the eruption of Mt. St. Helens has become a new metaphor for the very opposite of stability—for that greatest of twentieth century fears—cancer. Our uneasiness may well rest on more elusive levels than dirty windshields.

This comes far closer to home than anything else I've read about the "meaning" of the eruption, and yet for me it doesn't work. Maybe it would work better for men. The trouble is, I never saw St. Helens as a breast. Some mountains, yes: Twin Peaks in San Francisco, of course, and other round, sweet California hills—breasts, bellies, eggs, anything maternal, bounteous, yielding. But St. Helens in my eyes was never part of a woman; she is a woman. And not a mother but a sister.

These emotional perceptions and responses sound quite foolish when written out in rational prose, but the fact is that, to me, the eruption was all mixed up with the women's movement. It may be silly but there it is; along the same lines, do you know any woman who wasn't rooting for Genuine Risk to take the Triple Crown? Part of my satisfaction and exultation at each eruption was unmistakably feminist solidarity. You men think you're the only ones can make a really nasty mess? You think you got all the firepower, and God's on your side? You think you run things? Watch this, gents. Watch the Lady act like a woman.

For that's what she did. The well-behaved, quiet, pretty, serene, domestic creature peaceably yielding herself to the uses of man all of a sudden said NO. And she spat dirt and smoke and steam. She blackened half her face, in those first March days, like an angry brat. She fouled herself like a mad old harridan. She swore and belched and farted, threatened and shook and swelled, and then she spoke. They heard her voice two hundred miles away. Here I go, she said. I'm doing my thing now. Old Nobodaddy you better JUMP!

Her thing turns out to be more like childbirth than anything else, to my way of thinking. But not on our scale, not in our terms. Why should she speak in our terms or stoop to our scale? Why should she bear any birth that we can recognize? To us it is cataclysm and destruction and deformity. To her—well, for the language for it one must go to the scientists or to the poets. To the geologists. St. Helens is doing exactly what she "ought" to do—playing her part in the great pattern of events perceived by that noble discipline. Geology provides the only time-scale large enough to include the behavior of a volcano without deforming it. Geology, or poetry, which can see a mountain and a cloud as, after all, very similar phenomena. Shelley's cloud can speak for St. Helens:

> I silently laugh
> At my own cenotaph . . .
> And arise, and unbuild it again.

So many mornings waking I have seen her from the window before any other thing: dark against red daybreak, silvery in summer light, faint above river-valley fog. So many times I have watched her at evening, the faintest outline in mist, immense, remote, serene: the center, the central stone. A self across the air, a sister self, a stone. "The stone is at the center," I wrote in a poem about her years ago. But the poem is impertinent. All I can say is impertinent.

When I was writing the first draft of this essay in California, on July 23, she erupted again, sending her plume to 60,000 feet. Yesterday, August 7, as I was typing the words "the 'meaning' of the eruption," I checked out the study window and there it was, the towering blue cloud against the quiet northern sky—the fifth major eruption. How long may her labor be? A year, ten years, ten thousand? We cannot predict what she may or might or will do, now, or next, or for the rest of our lives, or ever. A threat: a terror: a fulfillment. This is what serenity is built on. This unmakes the metaphors. This is beyond us, and we must take it personally. This is the ground we walk on. ▪

5 On Mountains and Highlands

Dark as if cloven from darkness
were those mountains.

Night-angled fold on fold
they rose in mist and sunlight,

Their surging darkness
drums bells gongs imploring a god.

—Robert Hayden,
"An Inference of Mexico"

ALTHOUGH THE GREAT DIVERSITY AND DISTINCTIVENESS OF INDIGENOUS or traditional cultures resist generalization and stereotype, mountains and highlands have exhibited a sacred order that contributes to ritual, mythology, and oral tradition. As such, mountains have played primary roles in narratives on boundaries and the origin of things, places, and humans for many cultures.

Mountains have also played primary roles in western scientific narratives of a dynamic Earth. Several successive geologic theories have attempted to explain the existence of mountains: that they are the product of Earth's "skin" wrinkling as it contracted, rather like that of a dried apple or plum; that they are the products of a long-linear trough of thick sediments (a "geosyncline") that subsequently reversed its movements to produce mountains; and, most recently, that they are the product of plate tectonics. In the current thinking of geoscientists, mountains primarily form along plate margins where intense forces bend, break, displace, and uplift rocks.

The Earth now has three main belts of mountains: the Alpine-Himalayan belt stretching from Gibraltar to Southeast Asia; the Circum-Pacific belt that rims the margins of the Pacific Ocean; and the mountains of eastern Africa. The first belt is the product of convergence and collision of oceanic island arcs, small continental fragments, and finally Africa, Arabia, and pieces of Gondwanaland with Eurasia. This system includes the Himalaya (Sanskrit *hima* for snow and *alaya* for abode) as part of a 2,700-kilometer arc of the highest mountains on land (six to nine kilometers) that borders India and Tibet, the five-kilometer-high Tibetan Plateau, and the six-kilometer-high Pamir plateau in Tajikistan, Afghanistan, and Kazakhstan.

The Circum-Pacific belt is also associated with active convergent plate edges and in places with collision of island arcs or linear volcanic chains with each other, as in Indonesia, or with continents, as in Japan, Taiwan, Australia (New Guinea), western North America, and to some extent South America. The third belt is an area where the African continent has begun to come apart. The development of the East African Rifts and their uplift from equatorial jungle near sea level to present heights of two to six kilometers above sea level began about eight million years ago, simultaneously with the evolution of humans. Thus, in a sense, we *may* owe our existence as a species to the rise of mountains.

Remnants of older belts of deformed rocks resembling those in modern mountains, called *orogenic belts* (after the Greek *Oros* for mountain and *gen* for production of, birth) attest to the action of plate tectonics or some similar phenomena from the earliest times of Earth history. These ancient and fragmented belts include the Appalachian Mountains of eastern North America, the Caledonian Mountains of the British Isles, Scandinavia, Greenland, and Svalbard, the Ural Mountains of Russia, and the Tien Shan of China and central Asia, and the ancient deformed rocks of the centers of continents.

The pieces we have selected for this section give a small sampling of the vast literature on mountains. The chapter begins with an excerpt from the first volume of *Modern Painters* by John Ruskin (1819–1900), the great British art critic and social commentator of the Victorian Age, on the nature of mountains in the general structure of the Earth. In his view, "Mountains are, to the rest of the body of the earth, what violent muscular action is to the body of man." In the fall and early winter of 1873, Isabella Bird (1831–1904), an Englishwoman traveling alone by horseback, made an extended tour of the Colorado Rocky Mountains. What she called "no region for tourists and women" is now a popular resort and tourist destination. From *A Lady's Life in the Rocky Mountains* (1879–80), we include part of Bird's description of her climb up Long's Peak in the Colorado Front Range, only five years after its first known ascent. This narrative was originally written as a letter to her sister and later published as part of the book.

In "Basin and Range" in *Annals of the Former World,* John McPhee describes his cross-country excursion with Princeton geologist Ken Deffeyes from the ancient rifted lands of New Jersey to the Basin and Range province of Nevada and western Utah. The selection describes the mountains of this region, their formation by uplift and decay by erosion.

In "Look to the Mountaintop," Tewa poet and scholar Alfonso Ortiz (1939–97) considers the complex mosaic of overlapping tribal worlds defined by mountains in New Mexico and Arizona for the Pueblo and Yuman peoples, the Pima, the Navajos, and the Apaches.

For Native America, the case of the Black Hills is but one example of the direct connection of geology to historical issues of cultural conflict in this country. In 1980 the U.S. Supreme Court affirmed that Congress in 1877 had illegally taken the Black Hills from the Lakota for gold exploration and mining, and for opening the region to overland railroad routes. In the excerpt from *Land of the Spotted Eagle* (1933), Lakota elder Luther Standing Bear (1868?– 1939) describes some elements of the deep connection of his people to the *He Sapa* or Black Hills.

Ivan Doig (1939–) is a native of the Montana highlands and the only child of a ranch hand and ranch cook. A former ranch hand himself, newspaperman, and magazine editor with a Ph.D. in history, Doig writes of what he calls "that larger country: life." The highly acclaimed memoir *This House of Sky* was a finalist for the National Book Award. In an excerpt he describes a Montana mountainscape with detailed intimacy.

In *The Sacred Theory of the Earth,* Thomas Burnet (c. 1635–1715) tried to reconcile the accumulating observations of the Earth with the biblical account of Creation and the great Flood. He proposed that God's creation of an earthly paradise, an ordered world "smooth, regular, uniform," was destroyed by the Flood and left as wasted and broken landscapes. Mountains, as described here by Burnet, are part of the ruins of that world, "vast bodies thrown together in confusion."

Sir Archibald Geikie (1835–1924), a native of Edinburgh, Scotland, was a leading geologist of the British Isles and Europe. He served as director of the Scottish Geological Survey, professor of geology at the University of Edinburgh, and head of the Geological Survey of Great Britain, a post he held until his retirement. In addition to his scientific writings, Geikie wrote several well-known biographical and historical books. This excerpt from *Landscape in History and Other Essays* (1905) reflects on the highlands of Britain's Lake District and the poetry of William Wordsworth.

John Muir (1838–1914) was a naturalist, mountain man, outspoken proponent of wilderness preservation, and cofounder of the Sierra Club (in 1892). In his writings about regions he explored and tramped, Muir described and celebrated wildlands with eloquent passion, particularly the Sierra Nevada, which he first visited as a young man. *My First Summer in the Sierra* is based on journal entries for 1869. In it Muir describes "my forever memorable first High Sierra excursion, when I crossed the Range of Light, surely the brightest and best of all the Lord has built."

We conclude this section with the poem "Why" from Nanao Sakaki, self-proclaimed "unofficial examiner of the mountains and rivers of all Japan," poet, world traveler, and itinerant artist and scholar.

John Ruskin

FROM *Modern Painters*, Volume 1 (1843)

By truth of earth, we mean the faithful representation of the facts and forms of the bare ground, considered as entirely divested of vegetation, through whatever disguise, or under whatever modification the clothing of the landscape may occasion. Ground is to the landscape painter what the naked human body is to the historical. The growth of vegetation, the action of water and even of clouds upon it and around it, are so far subject and subordinate to its forms, as the folds of the dress and the fall of the hair are to the modulation of the animal anatomy. Nor is this anatomy always so concealed, but in all sublime compositions, whether of nature or art, it must be seen in its naked purity. The laws of the organization of the earth are distinct and fixed as those of the animal frame, simpler and broader, but equally authoritative and inviolable. Their results may be arrived at without knowledge of the interior mechanism; but for that very reason ignorance of them is the more disgraceful, and violation of them more unpardonable. They are in the landscape the foundation of all other truths, the most necessary, therefore, even if they were not in themselves attractive; but they are as beautiful as they are essential, and every abandonment of them by the artist must end in deformity as it begins in falsehood.
[. . .]
Mountains are, to the rest of the body of the earth, what violent muscular action is to the body of man. The muscles and tendons of its anatomy are, in the mountain, brought out with fierce and convulsive energy, full of expression, passion, and strength; the plains and the lower hills are the repose and the effortless motion of the frame, when its muscles lie dormant and concealed beneath the lines of its beauty, yet ruling those lines in their every undulation. This, then, is the first grand principle of the truth of the earth. The spirit of the hills is action, that of the lowlands repose; and between these there is to be found every variety of motion and of rest, from the inactive plain, sleeping like the firmament, with cities for stars, to the fiery peaks, which, with heaving bosoms and exulting limbs,

with the clouds drifting like hair from their bright foreheads, lift up their Titan hands to Heaven, saying, "I live for ever!"

But there is this difference between the action of the earth, and that of a living creature; that while the exerted limb marks its bones and tendons through the flesh, the excited earth casts off the flesh altogether, and its bones come out from beneath. Mountains are the bones of the earth, their highest peaks are invariably those parts of its anatomy which in the plains lie buried under five and twenty thousand feet of solid thickness of superincumbent soil, and which spring up in the mountain ranges in vast pyramids or wedges, flinging their garment of earth away from them on each side. The masses of the lower hills are laid over and against their sides, like the masses of lateral masonry against the skeleton arch of an unfinished bridge, except that they slope up to and lean against the central ridge: and finally, upon the slopes of these lower hills are strewed the level beds of sprinkled gravel, sand, and clay, which form the extent of the champaign. Here then is another grand principle of the truth of earth, that the mountains must come from under all, and be the support of all; and that everything also must be laid in their arms, heap above heap, the plains being the uppermost. ∎

Isabella Bird

FROM "Letter VII" in *A Lady's Life in the Rocky Mountains* (1879–80)

Estes Park, Colorado, October. [1873]

As this account of the ascent of Long's Peak could not be written at the time, I am much disinclined to write it, especially as no sort of description within my powers could enable another to realize the glorious sublimity, the majestic solitude, and the unspeakable awfulness and fascination of the scenes in which I spent Monday, Tuesday, and Wednesday.

Long's Peak, 14,700 feet high, blocks up one end of Estes Park, and dwarfs all the surrounding mountains. From it on this side rise, snow-born, the bright St. Vrain, and the Big and Little Thompson. By sunlight or moonlight its splintered grey crest is the one object which, in spite of wapiti and bighorn, skunk and grizzly, unfailingly arrests the eyes. From it come all storms of snow and wind, and the forked lightnings play around its head like a glory. It is one of the noblest of mountains, but in one's imagination it grows to be much more than a mountain. It becomes invested with a personality. In its caverns and abysses one comes to fancy that it generates and chains the strong winds, to let them loose in its fury. The thunder becomes its voice, and the lightnings do it homage. Other summits blush under the morning kiss of the sun, and turn pale the next moment; but it detains the first sunlight and holds it round its head for an hour at least, till it pleases to change from rosy red to deep blue; and the sunset, as if spell-bound, lingers latest on its crest. The soft winds which hardly rustle the pine needles down here are raging rudely up there round its motionless summit. The mark of fire is upon it; [. . .]

[On the mountain]

You know I have no head and no ankles, and never ought to dream of mountaineering; and had I known that the ascent was a real mountaineering feat I should not have felt the slightest ambition to perform it. As it is, I am only humiliated by my success, for "Jim" [mountain trapper and guide] dragged me up, like a bale of goods, by sheer force of muscle. At the

"Notch" the real business of the ascent began. Two thousand feet of solid rock towered above us, four thousand feet of broken rock shelved precipitously below; smooth granite ribs, with barely foothold, stood out here and there; melted snow refrozen several times, presented more serious obstacle; many of the rocks were loose, and tumbled down when touched. To me it was a time of extreme terror. I was roped to "Jim," but it was of no use; my feet were paralyzed and slipped on the bare rock, and he said it was useless to try to go that way, and we retraced our steps. I wanted to return to the "Notch," knowing that my incompetence would detain the party, and one of the young men said almost plainly that a woman was a dangerous encumbrance, but the trapper replied shortly that if it were not to take a lady up he would not go up at all. He went on to explore, and reported that further progress on the correct line of ascent was blocked by ice; and then for two hours we descended, lowering ourselves by our hands from rock to rock along a boulder-strewn sweep of 4,000 feet, patched with ice and snow, and perilous from rolling stones. My fatigue, giddiness, and pain from bruised ankles, and arms half pulled out of their sockets, were so great that I should never have gone halfway had not "Jim," *nolens volens*, dragged me along with a patience and skill, and withal a determination that I should ascend the Peak, which never failed. After descending about 2,000 feet to avoid the ice, we got into a deep ravine with inaccessible sides, partly filled with ice and snow and partly with large and small fragments of rock, which were constantly giving away, rendering footing very insecure. That part to me was two hours of painful and unwilling submission to the inevitable; of trembling, slipping, straining, of smooth ice appearing when it was least expected, and of weak entreaties to be left behind while the others went on. "Jim" always said that there was no danger, that there was only a short bad bit ahead, and that I should go up even if he carried me!

Slipping, faltering, gasping from the exhausting toil in the rarefied air, with throbbing hearts and panting lungs, we reached the top of the gorge and squeezed ourselves between two gigantic fragments of rock by a passage called the "Dog's Lift," when I climbed on the shoulders of one man and then was hauled up. This introduced us by an abrupt turn round the south-west angle of the Peak to a narrow shelf of considerable length, rugged, uneven, and so overhung by the cliff in some places that it is necessary to crouch to pass at all. Above, the Peak looks nearly vertical for 400 feet; and below, the most tremendous precipice I have ever seen descends in one unbroken fall. This is usually considered the most dangerous part of the ascent, but it does not seem so to me, for such foothold as there is is secure, and one fancies that it is possible to hold on with the hands. But there, and on the final, and, to my thinking, the worst part of the climb, one slip, and a breathing, thinking, human being would lie 3,000 feet below, a shapeless, bloody heap! "Ring" [the guide's dog] refused to traverse the Ledge, and remained at the "Lift" howling piteously.

From thence the view is more magnificent even than that from the "Notch." At the foot of the precipice below us lay a lovely lake, wood embosomed, from or near which the bright St. Vrain and other streams take their rise. I thought how their clear cold waters, growing turbid in the affluent flats, would heat under the tropic sun, and eventually form part of the great ocean river which renders our far-off islands habitable by impinging on their shores. Snowy ranges, one behind the other, extended to the distant horizon, folding in their wintry embrace the beauties of Middle Park. Pike's Peak, more than one hundred miles off, lifted that vast but shapeless summit which is the landmark of southern Colorado. There were snow patches, snow slashes, snow abysses, snow forlorn and soiled looking, snow pure and dazzling, snow glistening above the purple robe of pine worn by all the mountains; while away to the east, in limitless breadth, stretched the green-grey of the endless Plains. Giants everywhere reared their splintered crests. From thence, with a single sweep, the eye takes in a distance of 300 miles—that distance to the west, north, and south being made up of mountains ten, eleven, twelve, and thirteen thousand feet in height, dominated by Long's Peak, Gray's Peak, and Pike's Peak, all nearly the height of Mont Blanc! On the plains we traced the rivers by their fringe of cottonwoods to the distant Platte, and between us and them lay glories of mountain, canyon, and lake, sleeping in depths of blue and purple most ravishing to the eye.

As we crept from the ledge round a horn of rock I beheld what made me perfectly sick and dizzy to look at—the terminal Peak itself—a smooth, cracked face or wall of pink granite, as nearly perpendicular as anything could well be up which it was possible to climb, well deserving the name of the "American Matterhorn."

Scaling, not climbing, is the correct term for this last ascent. It took one hour to accomplish 500 feet, pausing for breath every minute or two. The only foothold was in narrow cracks or on minute projections on the granite. To get a toe in these cracks, or here and there in a scarcely obvious projection, while crawling on hands and knees, all the while tortured with thirst and gasping and struggling for breath, this was the climb; but at last the Peak was won. A grand, well-defined mountain top it is, a nearly level acre of boulders, with precipitous sides all round, the one we came up being the only accessible one. ▪

John McPhee

FROM "Basin and Range" in *Annals of the Former World* (1998)

Basin. Fault. Range. Basin. Fault. Range. A mile of relief between basin and range. Stillwater Range. Pleasant Valley. Tobin Range. Jersey Valley. Sonoma Range. Pumpernickel Valley. Shoshone Range. Reese River Valley. Pequop Mountains. Steptoe Valley. Ondographic rhythms of the Basin and Range. We are maybe forty miles off the interstate, in the Pleasant Valley basin, looking up at the Tobin Range. At the nine-thousand-foot level, there is a stratum of cloud against the shoulders of the mountains, hanging like a ring of Saturn. The summit of Mt. Tobin stands clear, above the cloud. When we crossed the range, we came through a ranch on the ridgeline where sheep were fenced around a running brook and bales of hay were bright green. Junipers in the mountains were thickly hung with berries, and the air was unadulterated gin. This country from afar is synopsized and dismissed as "desert"—the home of the coyote and the pocket mouse, the side-blotched lizard and the vagrant shrew, the MX rocket and the pallid bat. There are minks and river otters in the Basin and Range. There are deer and antelope, porcupines and cougars, pelicans, cormorants, and common loons. There are Bonaparte's gulls and marbled godwits, American coots and Virginia rails. Pheasants. Grouse. Sandhill cranes. Ferruginous hawks and flammulated owls. Snow geese. This Nevada terrain is not corrugated, like the folded Appalachians, like a tubal air mattress, like a rippled potato chip. This is not—in that compressive manner—a ridge-and-valley situation. Each range here is like a warship standing on its own, and the Great Basin is an ocean of loose sediment with these mountain ranges standing in it as if they were members of a fleet without precedent, assembled at Guam to assault Japan. Some of the ranges are forty miles long, others a hundred, a hundred and fifty. They point generally north. The basins that separate them—ten and fifteen miles wide—will run on for fifty, a hundred, two hundred and fifty miles with lone, daisy-petalled windmills standing over sage and wild rye. Animals tend to be content with their home ranges and not to venture out across the big dry valleys. "Imagine a chipmunk hiking across one of these basins," Deffeyes remarks. "The faunas in the high ranges here are quite distinct from one to another. Animals are isolated like Darwin's finches in the Galápagos. These ranges are truly islands."

Supreme over all is silence. Discounting the cry of the occasional bird, the wailing of a pack of coyotes, silence—a great spatial silence—is pure in the Basin and Range. It is a soundless immensity with mountains in it. You stand, as we do now, and look up at a high mountain front, and turn your head and look fifty miles down the valley, and there is utter silence. It is the silence of the winter forests of the Yukon, here carried high to the ridgelines of the ranges. "It is a soul-shattering silence," the physicist Freeman Dyson wrote of southern Nevada in *Disturbing the Universe*. "You hold your breath and hear absolutely nothing. No rustling of leaves in the wind, no rumbling of distant traffic, no chatter of birds or insects or children. You are alone with God in that silence. There in the white flat silence I began for the first time to feel a slight sense of shame for what we were proposing to do. Did we really intend to invade this silence with our trucks and bulldozers and after a few years leave it a radioactive junkyard?"

What Deffeyes finds pleasant here in Pleasant Valley is the aromatic sage. Deffeyes grew up all over the West, his father a petroleum engineer, and he says without apparent irony that the smell of sagebrush is one of two odors that will unfailingly bring upon him an attack of nostalgia, the other being the scent of an oil refinery. Flash floods have caused boulders the size of human heads to come tumbling off the range. With alluvial materials of finer size, they have piled up in fans at the edge of the basin. ("The cloudburst is the dominant sculptor here.") The fans are unconsolidated. In time to come, they will pile up to such enormous thicknesses that they will sink deep and be heated and compressed to form conglomerate. Erosion, which provides the material to build the fans, is tearing down the mountains even as they rise. Mountains are not somehow created whole and subsequently worn away. They wear down as they come up, and these mountains have been rising and eroding in fairly even ratio for millions of years—rising and shedding sediment steadily through time, always the same, never the same, like row upon row of fountains. In the

southern part of the province, in the Mojave, the ranges have stopped rising and are gradually wearing away. The Shadow Mountains. The Dead
Mountains, Old Dad Mountains, Cowhole Mountains, Bullion, Mule, and
Chocolate mountains. They are inselberge now, buried ever deeper in their
own waste. For the most part, though, the ranges are rising, and there can
be no doubt of it here, hundreds of miles north of the Mojave, for we are
looking at a new seismic scar that runs as far as we can see. It runs along
the foot of the mountains, along the fault where the basin meets the range.
From out in the valley, it looks like a long, buff-painted, essentially horizontal stripe. Up close, it is a gap in the vegetation, where plants growing
side by side were suddenly separated by several meters, where, one October
evening, the basin and the range—Pleasant Valley, Tobin Range—moved,
all in an instant, apart. They jumped sixteen feet. The erosion rate at
which the mountains were coming down was an inch a century. So in the
mountains' contest with erosion they gained in one moment about twenty
thousand years. These mountains do not rise like bread. They sit still for
a long time and build up tension, and then suddenly jump. Passively, they
are eroded for millennia, and then they jump again. They have been doing
this for about eight million years. This fault, which jumped in 1915,
opened like a zipper far up the valley, and, exploding into the silence, tore
along the mountain base for upward of twenty miles with a sound that suggested a runaway locomotive.

"This is the sort of place where you really do not put a nuclear plant,"
says Deffeyes. "There was other action in the neighborhood at the same
time—in the Stillwater Range, the Sonoma Range, Pumpernickel Valley.
Actually, this is not a particularly spectacular scarp. The lesson is that the
whole thing—the whole Basin and Range, or most of it—is alive. The earth
is moving. The faults are moving. There are hot springs all over the province.
There are young volcanic rocks. Fault scars everywhere. The world is split-
ting open and coming apart. You see a sudden break in the sage like this and
it says to you that a fault is there and a fault block is coming up. This is a
gorgeous, fresh, young, active fault scarp. It's growing. The range is lifting up.
The Nevada topography is what you see *during* mountain building. There
are no foothills. It is all too young. It is live country. This is the tectonic,

active, spreading, mountain-building world. To a nongeologist it's just ranges, ranges, ranges."

Most mountain ranges around the world are the result of compression, of segments of the earth's crust being brought together, bent, mashed, thrust and folded, squeezed up into the sky—the Himalaya, the Appalachians, the Alps, the Urals, the Andes. The ranges of the Basin and Range came up another way. The crust—in this region between the Rockies and the Sierra—is spreading out, being stretched, being thinned, being literally pulled to pieces. The sites of Reno and Salt Lake City, on opposite sides of the province, have moved apart sixty miles. The crust of the Great Basin has broken into blocks. The blocks are not, except for simplicity's sake, analogous to dominoes. They are irregular in shape. They more truly suggest stretch marks. Which they are. They trend nearly north-south because the direction of the stretching is roughly east-west. The breaks, or faults, between them are not vertical but dive into the earth at angles that average sixty degrees, and this, from the outset, affected the centers of gravity of the great blocks in a way that caused them to tilt. Classically, the high edge of one touched the low edge of another and formed a kind of trough, or basin. The high edge—sculpted, eroded, serrated by weather—turned into mountains. The detritus of the mountains rolled into the basin. The basin filled with water—at first, it was fresh blue water—and accepted layer upon layer of sediment from the mountains, accumulating weight, and thus unbalancing the block even further. Its tilt became more pronounced. In the manner of a seesaw, the high, mountain side of the block went higher and the low, basin side went lower until the block as a whole reached a state of precarious and temporary truce with God, physics, and mechanical and chemical erosion, not to mention, far below, the agitated mantle, which was running a temperature hotter than normal, and was, almost surely, controlling the action. Basin and range. Integral fault blocks: low side the basin, high side the range. For five hundred miles they nudged one another across the province of the Basin and Range. With extra faulting and whatnot, they took care of their own irregularities. Some had their high sides on the west, some on the east. The escarpment of the Wasatch Mountains—easternmost expression of this immense suite of mountains—faced west. The Sierra—the westernmost, the highest, the pre-dominant range, with Donner Pass only halfway up it—presented its escarpment to the east. As the developing Sierra made its skyward climb—as it went up past ten and twelve and fourteen thousand feet—it became so predominant that it cut off the incoming Pacific rain, cast a rain shadow (as the phenomenon is called) over lush, warm, Floridan and verdant Nevada. Cut it off and kept it dry. ▪

Alfonso Ortiz

FROM "Look to the Mountaintop" (1973)

One summer a few years ago an old man who was, like me, a Tewa Pueblo Indian, and I undertook a journey, a journey which in an allegorical sense reveals many of the uniquely American visions of life with which I shall deal in this essay. As such it may serve as a parable, if a true one, for all else to follow. We were driving to the country of the Utes in southwestern Colorado to share in the blessings of their Sun Dance. My companion had never in his life been in that part of Colorado before. As the massive outcropping that is known today as Chimney Rock loomed larger and larger beyond the road ahead, he became very alert. Pointing to it, he said, "There is Fire Mountain! It is just as the old people spoke of it." As he recognized distinct features of the place, he proceeded to unfold tale after tale of events in the early life of our people which took place at Fire Mountain and in the surrounding country. Every prominent feature along the road began to live for him, and as he spoke of that remembered place, we, each of us, began to realize that we were retracing a portion of the ancient journey of our people, a journey which began beneath a lake somewhere in this southwestern corner of Colorado who knows how many thousands of years ago, and a journey which, as long as there are Tewa to tell it, shall always end again at this lake of emergence.

So it was that by the time we neared the town of Pagosa Springs, it was no longer July 1963, but another time, a time in and out of time. This place is called Warm Sands in Tewa, for there are sands which are kept warm by the hot springs which gave birth to the town; sands which by themselves are said to be able to melt snow and moderate the mid-winter cold; sands for the obtaining of which our religious men in other times made winter pilgrimages. My companion and I both silently recalled ancestors who were among these religious men. He wanted to stop, to gather some of the warm sand from nearby springs, as did I. When we came to the sands, he knelt before the land, then he ran the sand through his fingers. And then he wept. He had never been here, but then he had never really left. He remembered his own grandfather and the other grandfathers who had preceded him here. He had heretofore never journeyed here, but now it was as if he had come home.

II

The notion "world view" denotes a distinctive vision of reality which not only interprets and orders the places and events in the experience of a people, but lends form, direction, and continuity to life as well. The preceding story well illustrates this, for implicit in it is a distinctive set of values, an identity, a feeling of rootedness, of belonging to a time and a place, and a felt sense of continuity with a tradition which transcends the experience of a single lifetime, a tradition which may be said to transcend even time. All of these are very closely bound up in Indian visions of life, whatever specific forms they may take. In this brief space I can only illustrate with the central notions shared by most of the tribes of the American Southwest, but so complex is this area that there are few ideas about what sort of life is most worth living found elsewhere in aboriginal North America, which are not also found here.

Among these widespread central notions, all of which I shall illustrate presently, is the belief that human life began, on a plant analogy, within the earth, usually beneath a lake or deep in a canyon. Another is that tribal space is defined by mountains which usually number four, one for each of the cardinal directions. Among the many people who subscribe to the belief that four mountains define tribal territory are the Navajos, all of the Pueblos, the Pima, and the Yuman tribes of the Gila River. Because these tribes are so numerous and once occupied contiguous territories in the two-state area of New Mexico and Arizona, we have a complex mosaic of overlapping tribal worlds defined by mountains. Variations of these beliefs extend into Central America where, some argue, they began among the great pre-Columbian civilizations which flourished there.

But mountains are more, much more, than boundary markers defining the tribal boundaries within which a people lives and carries on most of its meaningful, purposeful activities. The Pueblo peoples, for instance, believe that the four sacred mountains are pillars which hold up the sky and which divide the world into quarters. As such they are imbued with a high aura of mystery and sanctity. And this sacred meaning transcends all other meanings and functions. The Apaches, the most recent mountain dwellers among the southwestern Indians, believe that mountains are alive and the homes of supernaturals called "mountain people." They further believe that mountains are protectors from illness as well as external enemies, that they are the source of the powers of shamans as well as teachers of songs and other sacred knowledge to ordinary humans, and that, finally, mountains are defenders as well as definers of tribal territory. Indeed, the Chiracahua Apache believed that at the time of creation the goods of the earth were divided between Indians and whites, with Indians getting the mountains forevermore.

The Pima, a desert people occupying the great valley of the Gila and Salt Rivers in Arizona, stress, in addition, the harmful potential of the high mountains for those who have not ritually purified themselves prior to embarking on expeditions to these places and for those who stay too long. Mountain air, they assert, will cause a person's hair to turn white and make him age prematurely. Mountains will also cause one to have bad dreams and lose his strength. In Pima tradition the high mountains are not places where the young or ritually ignorant may safely trespass, and those who must go into the mountains must purify themselves ritually before and after each journey. At that they may stay only long enough to carry out a given task.

Both the Pima and their Yuman-speaking neighbors, the Maricopa, Quechan, and the Mojave, believe further that specific mountains and other high places are the homes of individual, named, supernatural teachers who may transport them there on dream journeys to give them specific kinds of knowledge and teach them specific skills. Here we have particular skills and attributes localized onto particular mountains and other high places inhabited by particular spirit patrons, spirit patrons which may be safely encountered only on dream journeys. I recall sitting in the sunshine one spring with an elder Maricopa when a large jet flew low over the reservation just outside of Phoenix, as it prepared to land. She looked up and remarked that whoever built the plane must have dreamed of the buzzard and visited his abode of a nearby mountain, for he who would learn how to fly can only do so from a buzzard.

The Pueblos and other farming Indians explicitly recognize, in addition, that mountains are also the sources of precious moisture and other life-sustaining blessings which may be appealed to and tapped by those who are eligible and who undergo ritual preparations. Viewed from this perspective, mountains are the destinations of regular pilgrimages, a purpose for which one of them, usually the one closest to a particular tribe, is singled out. Among many of the Pueblos only the medicine men are eligible to ascend to the very summits of the sacred mountains, and then only at dawn after they have shorn all their clothing and spent a whole night at the base praying and otherwise purifying themselves. The idea involved here—and it is one profoundly revealing of the meaning of the sacred mountains to the Pueblo people—is that one may approach the pure sacra or divinity represented by the mountaintop only after first attaining a state as close to pure innocence and purity as possible. We thus have the image of naked, unadorned, and undefiled man seeking blessings for his fellows by ascending to the summit of a mountain at dawn, with the sun at his back. Before ascending to the summit the pilgrims must also prepare an offering to the mountain and pray to it for permission to gather whatever it is they seek. These or similar conceptions are found all over aboriginal North America, wherever there are high places to be contemplated and climbed.

As sources of the ever-needed moisture, particular mountains may also be regarded as reliable predictors of the weather, even for a season yet months away. One Pueblo man was heard to remark of *Tsikomo*, the western sacred mountain of the Tewa: "*Tsikomo* is the chief of all the mountains because, depending on the kind of cloud patterns surrounding it at the end of July, we always know what kind of winter we shall have." This is not merely a romantic or mythopoeic notion, for moisture in the upper Rio Grande valley always comes from the west or north.

It is by now anticlimactic to say that sacred mountains may not be owned or fought over; they can belong to no one, so they cannot be subjects of dispute. *Tsikomo*, which is an eastern sacred mountain of the Navajo, was once the destination of regular pilgrimages by members of this tribe as well as by their enemies the Tewa. Yet the Tewa say that this is the one place they would never fight with the Navajo if they encountered them, however tempted they might be. And they quickly add that the Navajo felt the same way. This despite the fact that until mid-nineteenth century their enmity was so bitter that the Tewa came to single out the Navajo for special recognition as "our sacred enemies." The belief that sacred mountains cannot be owned is another which echoes over much of North America, with perhaps the best-known example being that of the Sioux conceptions surrounding the Black Hills of South Dakota, conceptions which are strikingly similar to those of the Tewa. ∎

Luther Standing Bear

FROM *Land of the Spotted Eagle* (1933)

Of all our domain, we loved, perhaps, the Black Hills the most. The Lakota had named these hills *He Sapa*, or Black Hills, on account of their color. The slopes and peaks were so heavily wooded with dark pines that from a distance the mountains actually looked black. In wooded recesses were numberless springs of pure water and numerous small lakes. There were wood and game in abundance and shelter from the storms of the plains. It was the favorite winter haunt of the buffalo and the Lakota as well. According to a tribal legend these hills were a reclining female figure from whose breasts flowed life-giving forces, and to them the Lakota went as a child to its mother's arms. The various entrances to the hills were very rough and rugged, but there was one very beautiful and easy pass through which both buffalo and Lakota entered the hills. This pass ran along a narrow stream bed which widened here and there but which in places narrowed so that the tall pines at the top of the cliffs arched their boughs, almost touching as they swayed in the winds. Every fall thousands of buffaloes and Lakotas went through this pass to spend the winter in the hills. *Pte ta tiyopa* it was called by the Lakotas, or "Gate of the buffalo." Today this beautiful pass is denuded of trees and to the white man it is merely "Buffalo Gap."

Two lovely legends of the Lakotas would be fine subjects for sculpturing—the Black Hills as the earth mother, and the story of the genesis of the tribe. Instead, the face of a white man is being outlined on the face of a stone cliff in the Black Hills. This beautiful region, of which the Lakota thought more than any other spot on earth, caused him the most pain and misery. These hills were to become prized by the white people for reasons far different from those of the Lakota. To the Lakota the magnificent forests and splendid herds were incomparable in value. To the white man everything was valueless except the gold in the hills. Toward the Indian the white people were absolutely devoid of sentiment, and when a people lack sentiment they are without compassion. So down went the Black Forest and to death went the last buffalo, noble animal and immemorial friend of the Lakota. As for the people who were native to the soil as the forests and buffalo—well, the gold-seekers did not understand them and never have. The white man will never know the horror and the utter bewilderment of the Lakota at the wanton destruction of the buffalo. What cruelty has not been glossed over with the white man's word—enterprise! If the Lakotas had been relinquishing any part of their territory voluntarily, the Black Hills would have been the last from the standpoint of traditional sentiment. So when by false treaties and trickery the Black Hills were forever lost, they were a broken people. The treaties, made supposedly to recompense them for the loss of this lovely region, were like all other treaties— worthless. But could the Lakota braves have foreseen the ignominy they were destined to endure, every man would have died fighting rather than give up his homeland to live in subjection and helplessness. ▪

Ivan Doig

FROM *This House of Sky* (1978)

It is not known just when in the 1860's the first white pioneers trickled into our area of south-central Montana, into what would come to be called the Smith River Valley. But if the earliest of them wagoned in on a day when the warm sage smell met the nose and the clear air lensed close the details of peaks two days' ride from there, what a glimpse into glory it must have seemed. Mountains stood up blue-and-white into the vigorous air. Closer slopes of timber offered the logs to hew homestead cabins from. Sage grouse nearly as large as hen turkeys whirred from their hiding places. And the expanse of it all: across a dozen miles and for almost forty along its bowed length, this home valley of the Smith River country lay open and still as a gray inland sea, held by buttes and long ridges at its northern and southern ends, and east and west by mountain ranges.

A new county had been declared here, bigger than some entire states in the East and vacant for the taking. More than vacant, evacuated: the Piegan Blackfeet tribes who had hunted across the land by then were pulling north, in a last ragged retreat to the long-grass prairies beyond the Missouri River. And promise of yet another sort: across on the opposite slopes of the Big Belt Mountains, placer camps around Helena were flushing gold out of every gravel gulch. With the Indians vanished and bonanza gold drawing in the town builders, how could this neighboring valley miss out on prosperity? No, unbridle imagination just for a moment, and it could not help but foretell all these seamless new miles into pasture and field, roads and a rail route, towns and homes.

Yet if they had had eyes for anything but the empty acres, those first-comers might have picked clues that this was a somewhat peculiar run of country, and maybe treacherous. Hints begin along the eastern skyline. There the Castle Mountains poke great turrets of stone out of black-green forest. From below in the valley, the spires look as if they had been engineered prettily up from the forest floor whenever someone took the notion, an entire mountain range of castle-builders' whims—until the fancy stone thrusts wore too thin in the wind and began to chink away, fissure by slow fissure. Here, if the valleycomers could have gauged it in some speedup of time, stood a measure of how wind and storm liked to work on that country, gladly nubbing down boulder if it stood in the way.

While the Castle Mountains, seen so in the long light of time, make a goblin horizon for the sun to rise over, the range to the west, the Big Belts, can cast some unease of its own on the valley. The highest peak of the range—penned into grandness on maps as Mount Edith, but always simply Old Baldy to those of us who lived with mountain upon mountain—thrusts up a bare summit with a giant crater gouged in its side. Even in hottest summer, snow lies in the great pock of crater like a patch on a gape of wound. Always, then, there is this reminder that before the time of men, unthinkable forces broke apart the face of the biggest landform the eye can find from any inch of the valley.

Nature's crankiness to the Big Belts did not quit there. The next summit to the south, Grass Mountain, grows its trees and grass in a pattern tipped upside down from every other mountain in sight. Instead of rising leisurely out of bunchgrass slopes which give way to timber reaching down from the crest, the Grassy is darkly cowled with timber at the bottom and opens into a wide generous pasture—a brow of prairie some few thousand feet higher than any prairie ought to be, all the length of its gentle summit.

Along the valley floor, omens still go on. The South Fork of the Smith River turns out to be little more than a creek named by an optimist. Or, rather, by some frontier diplomat, for as an early newspaperman explained in exactly the poetry the pawky little flow deserved, the naming took notice of a politician in the era of the Lewis and Clark expedition— *Secretary Smith of the Navy Department / The most progressive member of Jefferson's cabinet / . . . thus a great statesman, the expedition giver / is honored for all time in the name of "Smith River."* The overnamed subject of all that merely worms its way across the valley, generally kinking up three times the distance for every mile it flows and delivering all along the way more willow thickets and mud-browed banks than actual water. On the other hand, the water that is missing from the official streambed may arrive in some surprise gush somewhere else. A hot mineral pool erupting at an unnotable point in the valley gave the name to the county seat which built up around the steaming boil, White Sulphur Springs.

But whatever the quirks to be discovered in a careful look around, the valley and its walls of high country did fit that one firm notion the settlers held: empty country to fill up. Nor, in justice, could the eye alone furnish all that was vital to know. Probably it could not even be seen, at first, in the tides of livestock which the settlers soon were sending in seasonal flow between the valley and those curious mountains. What it took was experience of the climate, to remind you that those grazing herds of cattle and bands of sheep were not simply on the move into the mountains or back to the valley lowland. They were traveling between high country and higher, and in that unsparing landscape, the weather is rapidly uglier and more dangerous the farther up you go.

The country's arithmetic tells it. The very floor of the Smith River Valley rests one full mile above sea level. Many of the homesteads were set into the foothills hundreds of feet above that. The cold, storm-making mountains climb thousands of feet more into the clouds bellying over the Continental Divide to the west. Whatever the prospects might seem in a dreamy look around, the settlers were trying a slab of lofty country which often would be too cold and dry for their crops, too open to a killing winter for their cattle and sheep.

It might take a bad winter or a late rainless spring to bring out this fact, and the valley people did their best to live with calamity whenever it descended. But over time, the altitude and climate added up pitilessly, and even after a generation or so of trying the valley, a settling family might take account and find that the most plentiful things around them still were sagebrush and wind.

By the time I was a boy and Dad was trying in his own way to put together a life again, the doubt and defeat in the valley's history had tamped down into a single word. Anyone of Dad's generation always talked of a piece of land where some worn-out family eventually had lost to weather or market prices not as a farm or a ranch or even a homestead, but as a *place*. All those empty little clearings which ghosted that sage countryside—just the McLoughlin place there by that butte, the Vinton place over this ridge, the Kuhnes place, the Catlin place, the Winters place, the McReynolds place, all the tens of dozens of sites where families lit in the

valley or its rimming foothills, couldn't hold on, and drifted off. All of
them epitaphed with that barest of words, *place.*

One such place was where our own lives were compassed from. Southwest out of the valley into the most distant foothills of the Big Belts, both
the sage and the wind begin to grow lustier. Far off there, beyond the landmark rise called Black Butte and past even the long green pasture hump of
Grassy Mountain, a set of ruts can be found snaking away from the county
road. The track, worn bald by iron wagon-wheels and later by the hard tires
of Model T's, scuffs along red shale bluffs and up sagebrush gulches and past
trickling willow-choked creeks until at last it sidles across the bowed shoulder of a summit ridge. Off there in the abrupt openness, two miles and more
to a broad pitch of sage-soft slope, my father was born and grew up.

This sudden remote bowl of pasture is called the Tierney Basin—or
would be, if any human voice were there to say its name. Here, as far back
into the tumbled beginnings of the Big Belts as their wagons could go, a
double handful of Scots families homesteaded in the years just before this
century. Two deep Caledonian notions seem to have pulled them so far
into the hills: to raise sheep, and to graze them on mountain grass which
cost nothing.

A moment, cup your hands together and look down into them, and
there is a ready map of what these homesteading families had in mind. The
contours and life lines in your palms make the small gulches and creeks
angling into the center of the Basin. The main flow of water, Spring Creek,
drops down to squirt out there where the bases of your palms meet, the pass
called Spring Gulch. Toward these middle crinkles, the settlers clustered
in for sites close to water and, they hoped, under the wind. The braid of
lines, now, which runs square across between palms and wrists can be
Sixteenmile Creek, the canyoned flow which gives the entire rumpled
region its name—*the Sixteen country.* Thumbs and the upward curl of your
fingers represent the mountains and steep ridges all around. Cock the right
thumb a bit outward and it reigns as Wall Mountain does, prowing its rim-
rock out and over the hollowed land below. And on all that cupping rim of
unclaimed high country, the Scots families surely instructed one another
time and again, countless bands of sheep could find summer grass. ▪

Thomas Burnet

FROM *The Sacred Theory of the Earth* (1681–1689)

CHAPTER XI. *Concerning the Mountains of the Earth, their greatness and irregular Form, their Situation, Causes, and Origin.*

We have been in the hollows of the Earth, and the Chambers of the Deep, amongst the damps and steams of those lower Regions; let us now go air our selves on the tops of Mountains, where we shall have a more free and large Horizon, and quite another face of things will present it self to our observation.

The greatest objects of Nature are, methinks, the most pleasing to behold; and next to the great Concave of the Heavens, and those boundless Regions where the Stars inhabit, there is nothing that I look upon with more pleasure than the wide Sea and the Mountains of the Earth. There is something august and stately in the Air of these things, that inspires the mind with great thoughts and passions; We do naturally, upon such occasions, think of God and his greatness: and whatsoever hath but the shadow and appearance of INFINITE, as all things have that are too big for our comprehension, they fill and over-bear the mind with their Excess, and cast it into a pleasing kind of stupor and admiration.

And yet these Mountains we are speaking of, to confess the truth, are nothing but great ruines; but such as show a certain magnificence in Nature; as from old Temples and broken Amphitheaters of the *Romans* we collect the greatness of that people. But the grandeur of a Nation is less sensible to those that never see the remains and monuments they have left, and those who never see the mountainous parts of the Earth, scarce ever reflect upon the causes of them, or what power in Nature could be sufficient to produce them. The truth is, the generality of people have not sence and curiosity enough to raise a question concerning these things, or concerning the Original of them. You may tell them that Mountains grow out of the Earth like Fuzz-balls, or that there are Monsters under ground that throw up Mountains as Moles do Mole-hills; they will scarce raise one objection against your doctrine; or if you would appear more Learned, tell them that the Earth is a great Animal, and these are Wens that grow upon its body. This would pass current for Philosophy; so much is the World drown'd in stupidity and sensual pleasures, and so little inquisitive into the works of God and Nature.

There is nothing doth more awaken our thoughts or excite our minds to enquire into the causes of such things, than the actual view of them; as I have had experience my self when it was my fortune to cross the *Alps* and *Appennine* Mountains; for the sight of those wild, vast and indigested heaps of Stones and Earth, did so deeply strike my fancy, that I was not easie till I could give my self some tolerable account how that confusion came in Nature. 'Tis true, the height of Mountains compar'd with the Diameter of the Earth is not considerable, but the extent of them and the ground they stand upon, bears a considerable proportion to the surface of the Earth; and if from *Europe* we may take our measures for the rest, I eas-ily believe, that the Mountains do at least take up the tenth part of the dry land. [. . .]

'Tis certain that we naturally imagine the surface of the Earth much more regular than it is; for unless we be in some Mountainous parts, there seldom occur any great inequalities within so much compass of ground as we can, at once, reach with our Eye; and to conceive the rest, we multiply the same *Idea,* and extend it to those parts of the Earth that we do not see; and so fancy the whole Globe much more smooth and uniform than it is. But suppose a man was carri'd asleep out of a Plain Country, amongst the *Alps,* and left there upon the top of one of the highest Mountains, when he wak'd and look'd about him, he would think himself in an inchanted Country, or carri'd into another World; Every thing would appear to him so different to what he had ever seen or imagin'd before. To see on every hand of him a multitude of vast bodies thrown together in confusion, as those Mountains are; Rocks standing naked round about him; and the hollow Valleys gaping under him; and at his feet it may be, an heap of frozen Snow in the midst of Summer. He would hear the thunder come from below, and see the black Clouds hanging beneath him; Upon such a prospect, it would not be easie to him to perswade himself that he was still upon the same Earth; but if he did, he would be convinc'd, at least, that there are some Regions of it strangely rude, and ruine-like, and very different from what he had ever thought of before. But the inhabitants of these wild places are even with us; for those that live amongst the *Alps* and the great Mountains, think that all the rest of the Earth is like their Country, all broken into Mountains, and Valleys, and Precipices; They never see other, and most people think of nothing but what they have seen at one time or another. ■

Sir Archibald Geikie

FROM *Landscape in History and Other Essays* (1905)

One other mountainous district in Britain—that of the English Lakes—claims our attention for its influence on the progress of the national literature. Of all the isolated tracts of higher ground in these islands, that of the Lake District is the most eminently highland in character. It is divisible into two entirely distinct portions by a line drawn in a north-easterly direction from Duddon Sands to Shap Fells. South of that line the hills are comparatively low and featureless, though they enclose the largest of the lakes. They are there built up of ancient sedimentary strata, like those that form so much of the similar scenery in the uplands of Wales and the South of Scotland. But to the north of the line, most of the rocks are of a different nature, and have given rise to a totally distinct character of landscape. They consist of various volcanic materials which in early Palaeozoic time were piled up around submarine vents, and accumulated over the sea-floor to a thickness of many thousand feet. They were subsequently buried under the sediments that lie to the south, but, in after ages uplifted into land, their now diversified topography has been carved out of them by the meteoric agents of denudation. Thus pike and fell, crag and scar, mere and dale, owe their several forms to the varied degrees of resistance to the general waste offered by the ancient lavas and ashes. The upheaval of the district seems to have produced a dome-shaped elevation, culminating in a summit that lay somewhere between Helvellyn and Grasmere. At least from that center the several dales diverge, like the ribs from the top of a half-opened umbrella.

The mountainous tract of the Lakes, though it measures only some thirty-two miles from west to east by twenty-three from north to south, rises to heights of more than 3,000 feet, and as it springs almost directly from the margin of the Irish Sea, it loses none of the full effect of its elevation. Its fells present a thoroughly highland type of scenery, and have much of the dignity of far loftier mountains. Their sky-line often displays notched crests and rocky peaks, while their craggy sides have been carved into dark cliff-girt recesses, often filled with tarns, and into precipitous scars, which send long trails of purple scree down the grassy slopes.

Moreover, a mild climate and copious rainfall have tempered this natural asperity of surface by spreading over the lower parts of the fells and the bottoms of the dales a greener mantle than is to be seen among the mountains further north. Though the naked rock abundantly shows itself, it has been so widely draped with herbage and woodland as to combine the luxuriance of the lowlands with the near neighborhood of bare cliff and craggy scar.

Such was the scenery amidst which William Wordsworth was born and spent most of his long life. Thence he drew the inspiration which did so much to quicken the English poetry of the nineteenth century, and which has given to his dales and hills so cherished a place in our literature. The scenes familiar to him from infancy were loved by him to the end with an ardent and grateful affection which he never wearied of publishing to the world. No mountain-landscapes had ever before been drawn so fully, so accurately, and in such felicitous language. Every lineament of his hills and dales is depicted as luminously and faithfully in his verse as it is reflected on the placid surface of his beloved meres, but suffused by him with an ethereal glow of human sympathy. He drew from his mountain-landscape everything that

> "Can give an inward help, can purify
> And elevate, and harmonize and soothe."

It brought to him "authentic tidings of invisible things"; and filled him with

> "The sense
> Of majesty and beauty and repose,
> A blended holiness of earth and sky."

For his obligations to that native scenery he found continual expression.

> "Ye mountains and ye lakes,
> And sounding cataracts, ye mists and winds
> That dwell among the hills where I was born,

If in my youth I have been pure in heart,
If, mingling with the world, I am content
With my own modest pleasures, and have lived
With God and Nature communing, removed
From the little enmities and low desires—
The gift is yours."

Not only did his observant eye catch each variety of form, each passing
tint of color on his hills and valleys, he felt, as no poet before his time had
done, the might and majesty of the forces which, in the mountain-world,
we are shown how the surface of the world is continually modified.

"To him was given
Full many a glimpse of Nature's processes
Upon the exalted hills."

The thought of these glimpses led to one of the noblest outbursts in the
whole range of his poetry, where he gives way to the exuberance of his
delight in feeling himself, to use Byron's expression, "a portion of the
tempest"—

"To roam at large among unpeopled glens
And mountainous retirements, only trod
By devious footsteps; regions consecrate
To oldest time; and reckless of the storm,
 .   .   .   .   while the mists
Flying, and rainy vapours, call out shapes
And phantoms from the crags and solid earth,
 .   .   .   .   .   .   and while the streams
Descending from the region of the clouds,
And starting from the hollows of the earth,
More multitudinous every moment, rend
Their way before them—what a joy to roam
An equal among mightiest energies!"

In this passage Wordsworth seems to have had what he would have
called "a foretaste, a dim earnest" of that marvelous enlargement of the
charm and interest of scenery due to the process of modern science. When
he speaks of "regions consecrate to oldest time," he has a vague feeling

that somehow his glens and mountains belonged to a hoary antiquity, such as could be claimed by none of the verdant plains around. Had he written half a century later he would have enjoyed a clearer perception of the vastness of that antiquity and of the long succession of events with which it was crowded. ▪

John Muir

FROM *My First Summer in the Sierra* (1911)

July 26 [1869]. Ramble to the summit of Mount Hoffman, eleven thousand feet high, the highest point in life's journey my feet have yet touched. And what glorious landscapes are about me, new plants, new animals, new crystals, and multitudes of new mountains far higher than Hoffman, towering in glorious array along the axis of the range, serene, majestic, snow-laden, sun-drenched, vast domes and ridges shining below them, forests, lakes, and meadows in the hollows, the pure blue bell-flower sky brooding them all,—a glory day of admission into a new realm of wonders as if Nature had wooingly whispered, "Come higher." What questions I asked, and how little I know of all the vast show, and how eagerly, tremulously hopeful of some day knowing more, learning the meaning of these divine symbols crowded together on this wondrous page.

Mount Hoffman is the highest part of a ridge or spur about fourteen miles from the axis of the main range, perhaps a remnant brought into relief and isolated by unequal denudation. The southern slopes shed their water into Yosemite Valley by Tenaya and Dome Creeks, the northern in part into the Tuolumne River, but mostly into the Merced by Yosemite Creek. The rock is mostly granite, with some small piles and crests rising here and there in picturesque pillared and castellated remnants of red metamorphic slates. Both the granite and slates are divided by joints, making them separable into blocks like the stones of artificial masonry, suggesting the Scripture "He hath builded the mountains." Great banks of snow and ice are piled in hollows on the cool precipitous north side forming the highest perennial sources of Yosemite Creek. The southern slopes are much more gradual and accessible. Narrow slot-like gorges extend across the summit at right angles, which look like lanes, formed evidently by the erosion of less resisting beds. They are usually called "devil's slides," though they lie far above the region usually haunted by the devil; for though we read that he once climbed an exceeding high mountain, he cannot be much of a mountaineer, for his tracks are seldom seen above the timber-line.

The broad gray summit is barren and desolate-looking in general views, wasted by ages of gnawing storms; but looking at the surface in detail, one finds it covered by thousands and millions of charming plants with leaves and flowers so small they form no mass of color visible at a distance of a few hundred yards. [. . .]

How boundless the day seems as we revel in these storm-beaten sky gardens amid so vast a congregation of onlooking mountains! Strange and admirable it is that the more savage and chilly and storm-chafed the mountains, the finer the glow on their faces and the finer the plants they bear. The myriads of flowers tingeing the mountain-top do not seem to have grown out of the dry, rough gravel of disintegration, but rather they appear as visitors, a cloud of witnesses to Nature's love in what we in our timid ignorance and unbelief call howling desert. The surface of the ground, so dull and forbidding at first sight, besides being rich in plants, shines and sparkles with crystals: mica, hornblende, feldspar, quartz, tourmaline. The radiance in some places is so great as to be fairly dazzling, keen lance rays of every color flashing, sparkling in glorious abundance, joining the plants in their fine, brave beauty-work—every crystal, every flower a window opening into heaven, a mirror reflecting the Creator.

[. . .]

August 9. I went ahead of the flock, and crossed over the divide between the Merced and Tuolumne basins. The gap between the east end of the Hoffman spur and the mass of mountain rocks about Cathedral Peak, though roughened by ridges and waving folds, seems to be one of the channels of a broad ancient glacier that came from the mountains on the summit of the range. In crossing this divide the ice-river made an ascent of about five hundred feet from Tuolumne meadows. This entire region must have been overswept by ice.

From the top of the divide, and also from the big Tuolumne Meadows, the wonderful mountain called Cathedral Peak is in sight. From every point of view it shows marked individuality. It is a majestic temple of one stone, hewn from the living rock, and adorned with spires and pinnacles in regular cathedral style. The dwarf pines on the roof look like mosses. I hope some time to climb to it to say my prayers and hear the stone sermons.■

Nanao Sakaki

WHY (1981?)

Why climb a mountain?

Look! a mountain there.

I don't climb mountain.
Mountain climbs me.

Mountain is myself.
I climb on myself.

There is no mountain
          nor myself.
          Something
          moves up and down
          in the air. ▪

6 Rivers to the Sea

. . . suddenly he stood by the edge of a full-fed river.
Never in his life had he seen a river before—this sleek,
sinuous, full-bodied animal, chasing and chuckling,
gripping things with a gurgle and leaving them with
a laugh, to fling itself on fresh playmates that shook
themselves free, and were caught and held again.
All was a-shake and a-shiver—glints and gleams
and sparkles, rustle and swirl, chatter and bubble [. . .]
he sat on the bank, while the river still chattered on
to him, a babbling procession of the best stories in the
world, sent from the heart of the earth to be told at last
to the insatiable sea.

—Kenneth Grahame,
The Wind in the Willows

CONSIDER THIS: LIFE ON EARTH EVOLVED IN OCEANS. THE BODIES OF animals and plants consist largely of water. Moving water, from clouds to the sea, sculpts Earth's landscapes, from the highest mountains to flat lowlands. Rivers continue to be means and routes of human transport from continent interiors to the coasts. The work, patterns, and many forms of water play in human imagination and experience.

Two-thirds of Earth is covered by water. About 97 percent by volume of the planet's water is in oceans and is largely too saline for human use. The remaining 3 percent is fresh water, with almost all of it locked in ice caps and glaciers or buried deep beyond reach. Thus, only a small fraction of the total volume of water on the planet is easily available as soil moisture, usable groundwater, water vapor, lakes, and streams.

This freshwater supply is continuously collected, purified, recycled, and distributed in the sun-driven water (or hydrologic) cycle, one of the most basic and significant of Earth's features. In this cycle, water evaporates from the oceans, is carried as water vapor or clouds by Earth's great wind systems, then condenses and falls as precipitation. On land, it infiltrates into the ground and collects in small rivulets that come together to form small streams that cascade down steep canyons in mountains and join to form small rivers, which themselves join to form great rivers that carry the water back to the sea. Thus the cycle repeats itself over and over—all waters are one. As it does so, running water works to erode loosened rock material, depositing it in valleys and lakes, but ultimately carrying it to the sea.

Geologists speak of river systems, the network of stream channels and their shapes. The shapes of these systems and channels are a function of a dynamic equilibrium between the water's downcutting ability, which is related to the slope or steepness of the channel, the amount of sediment carried by the river, and the rate of uplift of the land. Consider, for example, the Mississippi River system with its extensive meandering channels, and its occasional large floods that overtop its banks to leave sediment on the floodplains beyond. Contrast that river with the Indus and Ganges rivers. These great river systems begin their journeys high in Tibet and India. They plunge steeply through canyons about as wide as the Grand Canyon of Arizona, and three times its depth, to the plains of Pakistan and India (and Bangladesh), across which they flow to the sea.

The selections in this chapter consider the universal, unifying cycle of water. The first five pieces recall watershed—a river's flow from source to end, the confluence of its tributaries—as sites of creation and change, as notions of time and the human body. We begin with "The Negro Speaks of Rivers" by Langston Hughes (1902–67). Hughes, whose work came to prominence during the Harlem Renaissance, writes of many rivers, including the Euphrates of the "Mesopotamian" fertile crescent.

Part of the Platte River essay "The Flow of the River" from Loren Eiseley's *The Immense Journey* (1957) follows. Eiseley once described this piece as a "concealed essay . . . in which personal anecdote was allowed gently to bring under observation thoughts of a more purely scientific nature." The Platte River is also of historical importance, as it was the main avenue of nineteenth-century Euro-American migration across the mid-continent to California and Oregon.

From Heaven Lake (1983) describes contemporary Indian poet and novelist Vikram Seth's hitchhiking journey from northwestern China to his home in Delhi, India, via Tibet and Nepal. The excerpt presents a day among waterfalls, as Seth walked from Nilamu to Zhangmu in Tibet after crossing the Himalaya along one of the steep wild rivers, in this case a tributary of the Ganges.

Wendell Berry, accomplished contemporary writer of nonfiction, poetry, and fiction, has examined human connections to the land and agriculture. In "A Native Hill," Berry writes with intimate detail gained by close, regular observation of the joining of streams. His is a literary description of the erosive work and ever-changing dynamics of streams.

From essayist, short-story writer, and naturalist-philosopher Barry Lopez's *River Notes* (1979), we include a section from "The Bend" in which the narrator attempts to "wrestle meaning" from a particular turn in a river, as if its quantification could provide some insight to a man's life.

An excerpt from John Wesley Powell's *The Exploration of the Colorado River and Its Canyons* (formerly *Canyons of the Colorado,* 1895) describes his river journey through the inner gorge of the Grand Canyon, first in 1869, then in 1871–72. Powell (1834–1902), as leader of one of the exploratory surveys, named the Colorado Plateau region and investigated the geology of the Southwest, ethnography of indigenous peoples, and land use of arid regions. He sought to understand large-scale landscape-shaping forces and, with his survey team (which included G. K. Gilbert), developed significant new ideas on the work of uplift and erosion, and the importance of time, and defined and advanced many basic principles and terms of geomorphology and structural geology.

The following pieces by Lucille Clifton and John McPhee tell of the Mississippi River. In the poem "the mississippi river empties into the gulf," Lucille Clifton ponders the meaning of the great circulation of water on Earth and considers the difference between geologic and human time scales.

In the chapter "Atchafalaya" from *The Control of Nature* (1989), John McPhee presents recent conflicts between the Mississippi River's meandering and flooding nature and the stationary nature of human settlement in the region. The selection considers the Old River Control, a comprehensive flood protection project for the lower Mississippi operated by the U.S. Army Corps of Engineers. To the Corps, the river has been an opponent that threatens commerce, homes, and the cities of Baton Rouge and New Orleans.

The last selection examines the marginal world where land and sea meet. Rachel Carson (1907–64), marine biologist and one of the most important environmental writers of the twentieth century, wrote *Under the Sea Wind* (1941), *The Sea Around Us* (1951), and the influential *Silent Spring* (1962), among other works. This selection from *The Edge of the Sea* (1955) describes the marginal world she explored in great detail throughout her life.

Langston Hughes

THE NEGRO SPEAKS OF RIVERS (1921)

I've known rivers:
I've known rivers ancient as the world and older than the
　　flow of human blood in human veins.

My soul has grown deep like the rivers.

I bathed in the Euphrates when dawns were young.
I built my hut near the Congo and it lulled me to sleep.
I looked upon the Nile and raised the pyramids above it.
I heard the singing of the Mississippi when Abe Lincoln
　　went down to New Orleans, and I've seen its muddy
　　bosom turn all golden in the sunset.

I've known rivers:
Ancient, dusky rivers.

My soul has grown deep like the rivers. ■

Loren Eiseley

FROM "The Flow of the River" in *The Immense Journey* (1957)

If there is magic on this planet, it is contained in water. Its least stir even, as now in a rain pond on a flat roof opposite my office, is enough to bring me searching to the window. A wind ripple may be translating itself into life. I have a constant feeling that some time I may witness that momentous miracle on a city roof, see life veritably and suddenly boiling out of a heap of rusted pipes and old television aerials. I marvel at how suddenly a water beetle has come and is submarining there in a spatter of green algae. Thin vapors, rust, wet tar and sun are an alembic remarkably like the mind; they throw off odorous shadows that threaten to take real shape when no one is looking.

Once in a lifetime, perhaps, one escapes the actual confines of the flesh. Once in a lifetime, if one is lucky, one so merges with sunlight and air and running water that whole eons, the eons that mountains and deserts know, might pass in a single afternoon without discomfort. The mind has sunk away into its beginnings among old roots and the obscure tricklings and movings that stir inanimate things. Like the charmed fairy circle into which man once stepped, and upon emergence learned that a whole century had passed in a single night, one can never quite define this secret; but it has something to do, I am sure, with common water. Its substance reaches everywhere; it touches past and prepares the future; it moves under the poles and wanders thinly in the heights of air. It can assume forms of exquisite perfection in a snowflake, or strip the living to a single shining bone cast up by the sea.

Many years ago, in the course of some scientific investigations in a remote western county, I experienced, by chance, precisely the sort of curious absorption by water—the extension of shape by osmosis—at which I have been hinting. You have probably never experienced in yourself the meandering roots of a whole watershed or felt your outstretched fingers touching, by some kind of clairvoyant extension, the brooks of snow-line glaciers at the same time that you were flowing toward the

Gulf over the eroded debris of worn-down mountains. A poet, MacKnight Black, has spoken of being "limbed . . . with waters gripping pole and pole." He had the idea, all right, and it is obvious that these sensations are not unique, but they are hard to come by; and the sort of extension of the senses that people will accept when they put their ear against a sea shell, they will smile at in the confessions of a bookish professor. What makes it worse is the fact that because of a traumatic experience in childhood, I am not a swimmer, and am inclined to be timid before any large body of water. Perhaps it was just this, in a way, that contributed to my experience.

As it leaves the Rockies and moves downward over the high plains towards the Missouri, the Platte River is a curious stream. In the spring floods, on occasion, it can be a mile-wide roaring torrent of destruction, gulping farms and bridges. Normally, however, it is a rambling, dispersed series of streamlets flowing erratically over great sand and gravel fans that are, in part, the remnants of a mightier Ice Age stream bed. Quicksand and shifting islands haunt its waters. Over it the prairie suns beat mercilessly throughout the summer. The Platte, "a mile wide and an inch deep," is a refuge for any heat-weary pilgrim along its shores. This is particularly true on the high plains before its long march by the cities begins.

The reason that I came upon it when I did, breaking through a willow thicket and stumbling out through ankle-deep water to a dune in the shade, is of no concern to this narrative. On various purposes of science I have ranged over a good bit of that country on foot, and I know the kinds of bones that come gurgling up through the gravel pumps, and the arrowheads of shining chalcedony that occasionally spill out of water-loosened sand. On that day, however, the sight of sky and willows and the weaving net of water murmuring a little in the shallows on its way to the Gulf stirred me, parched as I was with miles of walking, with a new idea: I was going to float. I was going to undergo a tremendous adventure.

The notion came to me, I suppose, by degrees. I had shed my clothes and was floundering pleasantly in a hole among some reeds when a great desire to stretch out and go with this gently insistent water began to pluck at me. Now to this bronzed, bold, modern generation, the struggle I waged with timidity while standing there in knee-deep water can only seem farcical;

yet actually for me it was not so. A near-drowning accident in childhood had scarred my reactions; in addition to the fact that I was a nonswimmer, this "inch-deep river" was treacherous with holes and quicksands. Death was not precisely infrequent along its wandering and illusory channels. Like all broad wastes of this kind, where neither water nor land quite prevails, its thickets were lonely and untraversed. A man in trouble would cry out in vain.

I thought of all this, standing quietly in the water, feeling the sand shifting away under my toes. Then I lay back in the floating position that left my face to the sky, and shoved off. The sky wheeled over me. For an instant, as I bobbed into the main channel, I had the sensation of sliding down the vast tilted face of the continent. It was then that I felt the cold needles of the alpine springs at my fingertips, and the warmth of the Gulf pulling me southward. Moving with me, leaving its taste upon my mouth and spouting under me in dancing springs of sand, was the immense body of the continent itself, flowing like the river was flowing, grain by grain, mountain by mountain, down to the sea. I was streaming over ancient sea beds thrust aloft where giant reptiles had once sported; I was wearing down the face of time and trundling cloud-wreathed ranges into oblivion. I touched my margins with the delicacy of a crayfish's antennae, and felt great fishes glide about their work.

I drifted by stranded timber cut by beaver in mountain fastnesses; I slid over shallows that had buried the broken axles of prairie schooners and the mired bones of mammoth. I was streaming alive through the hot and working ferment of the sun, or oozing secretively through shady thickets. I *was* water and the unspeakable alchemies that gestate and take shape in water, the slimy jellies that under enormous magnification of the sun writhe and whip upward as great barbeled fish mouths, or sink indistinctly back into the murk out of which they arose. Turtle and fish and the pinpoint chirpings of individual frogs are all watery projections, concentrations—as man himself is a concentration—of that indescribable and liquid brew which is compounded in varying proportions of salt and sun and time. It has appearances, but at its heart lies water, and as I was finally edged gently against a sand bar and dropped like any log, I tottered as I rose.

I knew once more the body's revolt against emergence into the harsh and unsupporting air, its reluctance to break contact with that mother element which still, at this late point in time, shelters and brings into being nine-enths of everything alive.

As for men, those myriad little detached ponds with their own swarming corpuscular life, what were they but a way that water has of going about beyond the reach of rivers? I, too, was a microcosm of pouring rivulets and floating driftwood gnawed by the mysterious animalcules of my own creation. I was three-fourths water, rising and subsiding according to the hollow knocking in my veins: a minute pulse like the eternal pulse that lifts Himalayas and which, in the following systole, will carry them away. ▪

Vikram Seth

FROM *From Heaven Lake* (1983)

This valley, draining southwards towards the Ganges, has eaten deeply into the Qinghai-Tibet plateau. It is thus surrounded on three sides by tall peaks. In fact the high terrace of Thong La through which we crossed the mountains before Nilamu lies considerably to the north of the main range of the Himalayas. From the high green walls of the Bhotakoshi, waterfalls drop hundreds of feet down to its surface: cascades and cataracts, twisting through plaited undergrowth or plunging straight down slabs of rock. The road folds back on itself or skirts around the edge of the hills in a continuous downward slope towards Zhangmu.

The tributaries of the river, when they are not actually waterfalls, rush down their angled gradient with an impetus terrifying for streams only a few meters across. The flood that struck here a few days ago has destroyed a small bridge on one such stream. Standing by the stream, its brown waters frothing a couple of steps away, its boulder-spattered spray striking my clothes and my face, I look down the thirty-degree slope that leads it to the Bhotakoshi a few hundred meters below. The stream roars through a gorge jammed with tree-trunks, rocks, debris. The walls of the gorge, ravaged by what must have been an enormous volume of water, show that the flood level at the height of the damage was thirty meters above even its present flow.

Every few minutes the troubled skin of the stream bursts open with new force against the rocks or the lashed logs which some roadworkers are trying to ease across it. A road-bridge here has been swept away, cutting Nilamu off from Zhangmu. At the moment there is only a makeshift bridge of logs, shaking on its rocky supports a few feet above the stream, with a cable on each side for a hand-grip. Though it is not more than five meters across, a slip would mean instant pulverization. I hesitate to cross with my unbalanced luggage, but there is no choice. Finally I gather courage from an eight-year-old Tibetan child who walks, nonchalantly swaying with the cables, over to the other side.

Along the main valley the devastation is apparent high above the Bhotakoshi's current flow. The firm concrete bridge between Nepal and China has disappeared, as has the checkpoint which used to be above it. Huge chunks of the road have been scooped away. Small huts, terraced fields clinging to the almost vertical walls of the gorge, standing crops, trees, rocks, everything was ground down to what villagers describe as a black slurry. The violence of the rains helped loose some of the compacted snows in the glaciers above, thus bloating the river with the stored precipitation of previous years. Here in Tibet only property was destroyed in the flood but lower down in Nepal a number of lives were lost, especially when the river swelled at night.

From far above, their source invisible above the green walls, the waterfalls empty themselves into the Bhotakoshi. Some, white, turbulent, resounding, scramble down the precipitous incline in short spasms of interrupted energy. Others plunge directly down, thundering into pools of rock and fern, bamboo, bramble and pine. One falls over the road itself, which crawls for a distance under a ledge.

I plan to be in Zhangmu before nightfall; there is time to pause along the way. I have still got some of Norbu's snacks; I eat them and an apple sitting at the end of a small spur. At this point I can count no less than eight waterfalls, bright silver in the noon light. It is difficult to imagine this view without them, and yet they are guests of just one season: when the monsoon passes they too will disappear.

"A land of streams! some, like a downward smoke,
Slow-dropping veils of thinnest lawn, did go—"

And across the valley against the grey verticality of a cliff a thin strand of water indeed vanishes into a mist or smoke atomized by the wind, to reappear, reconstituted from, it seems, the air itself into a liquid skein of light. There is enchantment in flowing water: I sit hypnotized by its beauty— water, the most unifying of the elements, that ties land and sea and air in one living ring. It has a channeled flow, unlike air, and its cycles are vaster, accepting all three states in nature. Snow and ice lie packed, perhaps for

years, on the crests and cwms, in glaciers, in the permanent snows, suddenly to crack and thaw and churn into the snowmelt. The water, deeper
than the highest peaks are high, lies in the sea until welled to the surface
and accepted into the air as vapor. The moist air circles the world to fall
again into the sea as snow or rain; or as snow or rain on the land.

"Highest good is like water," says Lao Tzu. "Because water excels in
benefiting the myriad creatures without contending with them and settles
where none would like to be, it comes close to the way." . . . "In the world
there is nothing more submissive and weak than water. Yet for attacking
that which is hard and strong nothing can surpass it." Tasteless, it accepts
all tastes, colorless, all colors, reflecting the sky, refracting the white
stones of its bed, dissolving or suspending the soils and minerals over
which it flows. The pulse of our bodies is liquid, as indeed all living pulses
are. Water dissolves the salt of the parable in the Upanishads, covers the
land of Genesis and flows by the paradise of the Koran. And the random
blur of noise, the tumult of light at which I now stare is the author of more
beauty even than itself: cirrus and cumulus, rainbow and storm cloud, the
strata of sunset, the indescribable scent of the first rains on the summer-
baked plains.

"It is all in the water": Scotch whiskey, Longjing tea. The universal element, it is yet so particular about its excellences. It "benefits the myriad
creatures," yet the vehement loveliness of the cataract is the cause of flood
and death in the overburdened stream below. Its substance yields to the
guiding rocks, yet its form outlives the rocks that direct and hinder its flow.

I will during my life be certain to drink some molecules of the water
passing this moment through the waterfall I see. Not only its image will
become a part of me; and its particles will become a part not merely of me
but of everyone in the world. The solid substances of the earth more easily cohere to particular people or nations, but those that flow—air, water—
are communal even within our lives.

With this curious thought, I gather up my luggage again and set off
down the road. Why, I wonder, do we stare so fixedly at water—at the sea,
at waterfalls, at streams? It seems perverse when the land is so much
more colorful, manifold, various. It is, I suppose, simply that water moves

while the land is static—or rather that its movements, the putting out of leaves, the movements of the earth's crust, are imperceptible to us. It is this visible movement of water, whether of the concentric ripples on a lake or of the "sounding cataract" falling whitely into chaos, that informs the purity of a uniform element with the varying impulse of life. ■

Wendell Berry

FROM "A Native Hill" (1981)

I have already passed the place where water began to flow in the little stream bed I am following. It broke into the light from beneath a rock ledge, a thin glittering stream. It lies beside me as I walk, overtaking me and going by, yet not moving, a thread of light and sound. And now from below comes the steady tumble and rush of the water of Camp Branch— whose nameless camp was it named for? —and gradually as I descend the sound of the smaller stream is lost in the sound of the larger.

The two hollows join, the line of the meeting of the two spaces obscured even in winter by the trees. But the two streams meet precisely as two roads. That is, the stream *beds* do; the one ends in the other. As for the meeting of the waters, there is no looking at that. The one flow does not end in the other, but continues in it, one with it, two clarities merged without a shadow.

All waters are one. This is a reach of the sea, flung like a net over the hill, and now drawn back to the sea. And as the sea is never raised in the earthly nets of fishermen, so the hill is never caught and pulled down by the watery net of the sea. But always a little of it is. Each of the gathering strands of the net carries back some of the hill melted in it. Sometimes, as now, it carries so little that the water seems to flow clear; sometimes it carries a lot and is brown and heavy with it. Whenever greedy or thoughtless men have lived on it, the hill has literally flowed out of their tracks into the bottom of the sea.

There appears to be a law that when creatures have reached the level of consciousness, as men have, they must become conscious of the creation; they must learn how they fit into it and what its needs are and what it requires of them, or else pay a terrible penalty: the spirit of the creation will go out of them, and they will become destructive; the very earth will depart from them and go where they cannot follow.

My mind is never empty or idle at the joinings of streams. Here is the work of the world going on. The creation is felt, alive and intent on its materials, in such places. In the angle of the meeting of the two streams stands the steep wooded point of the ridge, like the prow of an upturned boat—finished, as it was a thousand years ago, as it will be in a thousand years. Its becoming is only incidental to its being. It will be because it is. It has no aim or end except to be. By being, it is growing and wearing into what it will be. The fork of the stream lies at the foot of the slope like hammer and chisel laid down at the foot of a finished sculpture. But the stream is no dead tool; it is alive, it is still at its work. Put your hand to it to learn the health of this part of the world. It is the wrist of the hill.

Perhaps it is to prepare to hear some day the music of the spheres that I am always turning my ears to the music of streams. There is indeed a music in streams, but it is not for the hurried. It has to be loitered by and imagined. Or imagined *toward,* for it is hardly for men at all. Nature has a patient ear. To her the slowest funeral march sounds like a jig. She is satisfied to have the notes drawn out to the lengths of days or weeks or months. Small variations are acceptable to her, modulations as leisurely as the opening of a flower.

The stream is full of stops and gates. Here it has piled up rocks in its path, and pours over them into a tiny pool it has scooped at the foot of its fall. Here it has been dammed by a mat of leaves caught behind a fallen limb. Here it must force a narrow passage, here a wider one. Tomorrow the flow may increase or slacken, and the tone will shift. In an hour or a week that rock may give way, and the composition will advance by another note. Some idea of it may be got by walking slowly along and noting the changes as one passes from one little fall or rapid to another. But this is a highly simplified and diluted version of the real thing, which is too complex and widespread ever to be actually heard by us. The ear must imagine an impossible patience in order to grasp even the unimaginableness of such music.

But the creation is musical, and this is a part of its music, as bird song is, or the words of poets. The music of the streams is the music of the shaping of the earth, by which rocks are pushed and shifted downward toward the level of the sea. ■

Barry Lopez

FROM "The Bend" in *River Notes* (1979)

In the evenings I walk down and stand in the trees, in light paused just so in the leaves, as if the change in the river here were not simply known to me but apprehended. It did not start out this way; I began with the worst sort of ignorance, the grossest inquiries. Now I ask very little. I observe the swift movement of water through the nation of fish at my feet. I wonder privately if there are for them, as there are for me, moments of faith.

The river comes around from the southeast to the east at this point: a clean shift of direction, water deep and fast on the outside of the curve, flowing slower over the lip of a broad gravel bar on the inside, continuing into a field of shattered boulders to the west.

I kneel and slip my hands like frogs beneath the surface of the water. I feel the wearing away of the outer edge, the exposure of rootlets, the undermining. I imagine eyes in the tips of my fingers, like the eyestalks of crayfish. Fish stare at my fingertips and bolt into the river's darkness. I withdraw my hands, conscious of the trespass. The thought that I might be observed disturbs me.

I've wanted to take the measure of this turn in the river, grasp it for private reasons. I feel closer to it now. I know which deer drink at which spots on this bank. I know of the small screech owl nesting opposite. (I would point him out to you by throwing a stone in that direction but the gesture would not be appropriate.) I am familiar with the raccoon and fisher whose tracks appear here, can even tell them apart in the dark by delicately fingering the rim of their prints in the soil. I can hear the preparations of muskrats. On cold, damp nights I am aware of the fog of birds' breath that rolls oceanic through the trees above. Out there, I know which rocks are gripped by slumbering water striders, and where beneath the water lie the slipcase homes of caddis fly larvae.

I feel I am coming closer to it.

For myself, each day more of me slips away. Absorbed in seeing how the water comes through the bend, just so, I am myself, sliding off.

The attempt to wrestle meaning from this spot began poorly, with illness. A pain, slow in coming like so many, that seemed centered in the back of my neck. Then an acute yearning, as strong as the wish to be loved, pain along the ribs, and my legs started to give way. I awoke in the morning with my hands over my face as though astonished by my own dreams. As the weeks went on I moved about less and less, until finally I went to bed and lay there like summer leaves. I could hear the rain in the woods in the afternoon; the sound of the river, like the laughing of horses; smell faintly through the open window the breath of bears. Between these points I was contained, closed off like a spider by the design of a web. I tried to imagine that I was well, but the points of my imagination impaled me, and then a sense of betrayal emptied me.

I began to think (as on a staircase descending to an unexamined basement) about the turn in the river. If I could understand this smoothly done change of direction I could imitate it, I reasoned, just as a man puts what he reads in a story to use, substituting one point for another as he needs.

Several things might be measured I speculated: the rate of flow of the water, the erosion of the outer bank, the slope of the adjacent mountains, the changing radius of curvature as the river turned west. It could be revealed neatly, affirmed with graphic authority.

I became obsessed with its calculation. I lay the plan out first in my head, without recourse to paper. The curve required calculus, and so some loss of accuracy; and the precise depth of the river changed from moment to moment, as did its width. But I could abide this for the promise of insight into my life.

I called on surveyors, geodesic scientists, hydrologists. It was the work of half a year. It involved them in the arduous toting of instruments back and forth across the river and in tedious calculation. I asked that exacting journals be kept, that no scrap of description be lost. There were arguments, of course. I required that renderings be done again, over and over. I became convinced that in this wealth of detail a fixed reason for the river's graceful turn would inevitably be revealed.

The workmen, defeated by the precision required, in an anger all their own, hurled their theodolites into the trees. (The repair of these instruments consumed more time.) I understood that fights broke out. But I saw none of this. I lay alone in the room and those in my employ came and went politely with their notes. I knew they thought it pointless, but there was their own employment to be considered, and they said the wage was fair.

Finally they reduced the bend in the river to an elegant series of equations, and the books containing them and a bewildering list of variables were all gathered together and brought to my room. I had them placed on the floor, stacked in a corner. I suddenly had the strength for the first time, staring at this pile, to move, but I was afraid. I put it off until morning; I felt my recovery was certain, believed even more forcefully now that my own resolution was at hand by an incontestable analogy.

That night I awoke to hear the dripping of water. From the direction of the pile of notes came the sound of mergansers, the explosive sound they make when they are surprised on the water and suddenly fly off.

I lay back.

Moss grew eventually on the books. They began after a while to harden, to resemble the gray boulders in the river. Years passed. I smelled cottonwood on spring afternoons, and would imagine sunshine crinkling on the surface of the water.

In winter the windows remained open because I could not reach them.

One morning, without warning, I came to a dead space in my depression, a sudden horizontal view, which I seized. I pried myself from the room, coming down the stairs slow step by step, all the while calling out. Bears heard me (or were already waiting at the door). I told them I needed to be near the river. They carried me through the trees (growling, for they are not used to working together), throwing their shoulders to the alders until we stood at the outer bank.

Then they departed, leaving the odor of bruised grass and cracked bone hanging in the air.

The first thing I did was to feel, raccoonlike, with the tips of my fingers the soil of the bank just below the water's edge. I listened for the sound of water on the outer bar. I observed the hunt of the caddis fly.

I am now taking measure of the bend in these experiences.

I have lost, as I have said, some sense of myself. I no longer require much. And though I am hopeful of recovery, an adjustment as smooth as the way the river lies against the earth at this point, this is no longer the issue with me. I am more interested in this: from above, to a hawk, the bend must appear only natural and I for the moment inseparably a part, like salmon or a flower. I cannot say well enough how this single perception has dismantled my loneliness. ∎

John Wesley Powell

FROM *The Exploration of the Colorado River and Its Canyons*
    (formerly *Canyons of the Colorado,* 1895)

CHAPTER XI.   *From the Little Colorado to the Foot of the Grand Canyon*

August 13.—We are now ready to start on our way down the Great
Unknown. Our boats, tied to a common stake, chafe each other as they are
tossed by the fretful river. They ride high and buoyant, for their loads are
lighter than we could desire. We have but a month's rations remaining. The
flour has been resifted through the mosquito-net sieve; the spoiled bacon has
been dried and the worst of it boiled; the few pounds of dried apples have
been spread in the sun and reshrunken to their normal bulk. The sugar has
all melted and gone on its way down the river. But we have a large sack of
coffee. The lightening of the boats has this advantage: they will ride the
waves better and we shall have but little to carry when we make a portage.

We are three quarters of a mile in the depths of the earth, and the great
river shrinks into insignificance as it dashes its angry waves against the
walls and cliffs that rise to the world above; the waves are but puny ripples,
and we but pigmies, running up and down the sands or lost among the
boulders.

We have an unknown distance yet to run, an unknown river to explore.
What falls there are, we know not; what rocks beset the channel, we know
not; what walls rise over the river, we know not. Ah, well! We may conjecture many things. The men talk as cheerfully as ever; jests are bandied
about freely this morning; but to me the cheer is somber and the jests are
ghastly.

With some eagerness and some anxiety and some misgiving we enter
the canyon below and are carried along by the swift water through walls
which rise from its very edge. They have the same structure that we
noticed yesterday—tiers of irregular shelves below, and, above these, steep
slopes to the foot of marble cliffs. We run six miles in a little more than
half an hour and emerge into a more open portion of the canyon, where
high hills and ledges of rock intervene between the river and the distant

walls. Just at the head of this open place the river runs across a dike; that is, a fissure in the rocks, open to depths below, was filled with eruptive matter, and this on cooling was harder than the rocks through which the crevice was made, and when these were washed away the harder volcanic matter remained as a wall, and the river has cut a gateway through it several hundred feet high and as many wide. As it crosses the wall, there is a fall below and a bad rapid, filled with boulders of trap; so we stop to make a portage. Then on we go, gliding by hills and ledges, with distant walls in view; sweeping past sharp angles of rock; stopping at a few points to examine rapids, which we find can be run, until we have made another five miles, when we land for dinner.

Then we let down with lines over a long rapid and start again. Once more the walls close in, and we find ourselves in a narrow gorge, the water again filling the channel and being very swift. With great care and constant watchfulness we proceed, making about four miles this afternoon, and camp in a cave.

August 14.—At daybreak we walk down the bank of the river, on a little sandy beach, to take a view of a new feature in the canyon. Heretofore hard rocks have given us bad river; soft rocks, smooth water; and a series of rocks harder than any we have experienced sets in. The river enters the gneiss! We can see but a little way into the granite gorge, but it looks threatening.

After breakfast we enter on the waves. At the very introduction it inspires awe. The canyon is narrower than we have ever before seen it; the water is swifter; there are but few broken rocks in the channel; but the walls are set, on either side, with pinnacles and crags; and sharp, angular buttresses, bristling with wind- and wave-polished spires, extend far out into the river.

Ledges of rock jut into the stream, their tops sometimes just below the surface, sometimes rising a few or many feet above; and island ledges and island pinnacles and island towers break the swift course of the stream into chutes and eddies and whirlpools. We soon reach a place where a creek comes in from the left, and, just below, the channel is choked with boulders, which have washed down this lateral canyon and formed a dam, over which there is a fall of 30 or 40 feet; but on the boulders foothold can be had, and we make a portage. Three more such dams are found. Over one we make a portage; at the other two are chutes through which we can run.

As we proceed the granite rises higher, until nearly a thousand feet of the lower part of the walls are composed of this rock.

About eleven o'clock we hear a great roar ahead, and approach it very cautiously. The sound grows louder and louder as we run, and at last we find ourselves above a long, broken fall, with ledges and pinnacles of rock obstructing the river. There is a descent of perhaps 75 or 80 feet in a third of a mile, and the rushing waters break into great waves on the rocks, and lash themselves into a mad, white foam. We can land just above, but there is no foothold on either side by which we can make a portage. It is nearly a thousand feet to the top of the granite; so it will be impossible to carry our boats around, though we can climb to the summit up a side gulch and, passing along a mile or two, descend to the river. This we find on examination; but such a portage would be impracticable for us, and we must run the rapid or abandon the river. There is no hesitation. We step into our boats, push off, and away we go, first on smooth but swift water, then we strike a glassy wave and ride to its top, down again into the trough, up again on a higher wave, and down and up on waves higher and still higher until we strike one just as it curls back, and a breaker rolls over our little boat. Still on we speed, shooting past projecting rocks, till the little boat is caught in a whirlpool and spun round several times. At last we pull out again into the stream. And now the other boats have passed us. The open compartment of the "Emma Dean" is filled with water and every breaker rolls over us. Hurled back from a rock, now on this side, now on that, we are carried into an eddy, in which we struggle for a few minutes, and are then out again, the breakers still rolling over us. Our boat is unmanageable, but she cannot sink, and we drift down another hundred yards through breakers—how, we scarcely know. We find the other boats have turned into an eddy at the foot of the fall and are waiting to catch us as we come, for the men have seen that our boat is swamped. They push out as we come near and pull us in against the wall. Our boat bailed, on we go again.

The walls now are more than a mile in height—a vertical distance difficult to appreciate. Stand on the south steps of the Treasury building in Washington and look down Pennsylvania Avenue to the Capitol; measure this distance overhead, and imagine cliffs to extend to that altitude, and you will understand what is meant; or stand at Canal Street in New York and look up Broadway to Grace Church, and you have about the distance; or stand at Lake Street bridge in Chicago and look down to the Central Depot, and you have it again.

A thousand feet of this is up through granite crags; then steep slopes and perpendicular cliffs rise one above another to the summit. The gorge is black and narrow below, red and gray and flaring above, with crags and angular projections on the walls, which, cut in many places by side canyons, seem to be a vast wilderness of rocks. Down in these grand, gloomy depths we glide, ever listening, for the mad waters keep up their roar; ever watching, ever peering ahead, for the narrow canyon is winding and the river is closed in so that we can see but a few hundred yards, and what there may be below we know not; so we listen for falls and watch for rocks, stopping now and then in the bay of a recess to admire the gigantic scenery; and ever as we go there is some new pinnacle or tower, some crag or peak, some distant view of the upper plateau, some strangely shaped rock, or some deep, narrow side canyon.

Then we come to another broken fall, which appears more difficult than the one we ran this morning. A small creek comes in on the right, and the first fall of the water is over boulders, which have been carried down by this lateral stream. We land at its mouth and stop for an hour or two to examine the fall. It seems possible to let down with lines, at least a part of the way, from point to point, along the right-hand wall. So we make a portage over the first rocks and find footing on some boulders below. Then we let down one of the boats to the end of her line, when she reaches a corner of the projecting rock, to which one of the men clings and steadies her while I examine an eddy below. I think we can pass the other boats down by us and catch them in the eddy. This is soon done, and the men in the boats in the eddy pull us to their side. On the shore of this little eddy there is about two feet of gravel beach above the water. Standing on this beach, some of the men take the line of the little boat and let it drift down against another projecting angle. Here is a little shelf, on which a man from my boat climbs, and a shorter line is passed to him, and he fastens the boat to the side of the cliff; then the second one is let down, bringing the line of the third. When the second boat is tied up, the two men standing on the beach above spring into the last boat, which is pulled up alongside of ours; then we let down the boats for 25 or 30 yards by walking along the shelf, landing them again in the mouth of a side canyon. Just below this there is another pile of boulders, over which we make another portage. From the foot of these rocks we can climb to another shelf, 40 or 50 feet above the water.

On this bench we camp for the night. It is raining hard, and we have no shelter, but find a few sticks which have lodged in the rocks, and kindle a fire and have supper. We sit on the rocks all night, wrapped in our *ponchos*, getting what sleep we can.

August 15.—This morning we find we can let down for 300 or 400 yards, and it is managed in this way: we pass along the wall by climbing from projecting point to point, sometimes near the water's edge, at other places 50 or 60 feet above, and hold the boat with a line while two men remain aboard and prevent her from being dashed against the rocks and keep the line from getting caught on the wall. In two hours we have brought them all down, as far as it is possible, in this way. A few yards below, the river strikes with great violence against a projecting rock and our boats are pulled up in a little bay above. We must now manage to pull out of this and clear the point below. The little boat is held by the bow obliquely up the stream. We jump in and pull out only a few strokes, and sweep clear of the dangerous rock. The other boats follow in the same manner and the rapid is passed.

It is not easy to describe the labor of such navigation. We must prevent the waves from dashing the boats against the cliffs. Sometimes, where the river is swift, we must put a bight of rope about a rock, to prevent the boat from being snatched from us by a wave; but where the plunge is too great or the chute too swift, we must let her leap and catch her below or the undertow will drag her under the falling water and sink her. Where we wish to run her out a little way from shore through a channel between rocks, we first throw in little sticks of driftwood and watch their course, to see where we must steer so that she will pass the channel in safety. And so we hold, and let go, and pull, and lift, and ward—among rocks, around rocks, and over rocks.

And now we go on through this solemn, mysterious way. The river is very deep, the canyon very narrow, and still obstructed, so that there is no steady flow of the stream; but the waters reel and roll and boil, and we are scarcely able to determine where we can go. Now the boat is carried to the right, perhaps close to the wall; again, she is shot into the stream, and perhaps is dragged over to the other side, where, caught in a whirlpool, she spins about. We can neither land nor run as we please. The boats are entirely unmanageable; no order in their running can be preserved; now one, now another, is ahead, each crew laboring for its own preservation. In such a place we come to another rapid. Two of the boats run it perforce. One succeeds in landing, but there is no foothold by which to make a portage and she is pushed out again into the stream. The next minute a great reflex wave fills the open compartment; she is water-logged, and drifts unmanageable. Breaker after breaker rolls over her and one capsizes her. The men are thrown out; but they cling to the boat, and she drifts down some distance alongside of us and we are able to catch her. She is soon bailed out and the men are aboard once more; but the oars are lost, and so a pair from the "Emma Dean" is spared. Then for two miles we find smooth water.

Clouds are playing in the canyon to-day. Sometimes they roll down in great masses, filling the gorge with gloom; sometimes they hang aloft from wall to wall and cover the canyon with a roof of impending storm, and we can peer long distances up and down this canyon corridor, with its cloud-roof overhead, its walls of black granite, and its river bright with the sheen of broken waters. Then a gust of wind sweeps down a side gulch and, making a rift in the clouds, reveals the blue heavens, and a stream of sunlight pours in. Then the clouds drift away into the distance, and hang around crags and peaks and pinnacles and towers and walls, and cover them with a mantle that lifts from time to time and sets them all in sharp relief. Then baby clouds creep out of side canyons, glide around points, and creep back again into more distant gorges. Then clouds arrange in strata across the canyon, with intervening vista views to cliffs and rocks beyond.

The clouds are children of the heavens, and when they play among the rocks they lift them to the region above.

It rains! Rapidly little rills are formed above, and these soon grow into brooks, and the brooks grow into creeks and tumble over the walls in innumerable cascades, adding their wild music to the roar of the river. When the rain ceases the rills, brooks, and creeks run dry. The waters that fall during a rain on these steep rocks are gathered at once into the river; they could scarcely be poured in more suddenly if some vast spout ran from the clouds to the stream itself. When a storm bursts over the canyon a side gulch is dangerous, for a sudden flood may come, and the inpouring waters will raise the river so as to hide the rocks.

Early in the afternoon we discover a stream entering from the north—a clear, beautiful creek, coming down through a gorgeous red canyon. We land and camp on a sand beach above its mouth, under a great, over-spreading tree with willow-shaped leaves. ▪

Lucille Clifton

FROM *The Terrible Stories* (1996)

THE MISSISSIPPI RIVER EMPTIES INTO THE GULF

and the gulf enters the sea and so forth,
none of them emptying anything,
all of them carrying yesterday
forever on their white tipped backs,
all of them dragging forward tomorrow.
it is the great circulation
of the earth's body, like the blood
of the gods, this river in which the past
is always flowing.   every water
is the same water coming round.
everyday someone is standing on the edge
of this river, staring into time,
whispering mistakenly:
only here. only now.   ▪

John McPhee

FROM "Atchafalaya" in *The Control of Nature* (1989)

The Mississippi River, with its sand and silt, has created most of Louisiana, and it could not have done so by remaining in one channel. If it had, southern Louisiana would be a long narrow peninsula reaching into the Gulf of Mexico. Southern Louisiana exists in its present form because the Mississippi River has jumped here and there within an arc about two hundred miles wide, like a pianist playing with one hand—frequently and radically changing course, surging over the left or the right bank to go off in utterly new directions. Always it is the river's purpose to get to the Gulf by the shortest and steepest gradient. As the mouth advances southward and the river lengthens, the gradient declines, the current slows, and sediment builds up the bed. Eventually, it builds up so much that the river spills to one side. Major shifts of that nature have tended to occur roughly once a millennium. The Mississippi's main channel of three thousand years ago is now the quiet water of Bayou Teche, which mimics the shape of the Mississippi. Along Bayou Teche, on the high ground of ancient natural levees, are Jeanerette, Breaux Bridge, Broussard, Olivier—arcuate strings of Cajun towns. Eight hundred years before the birth of Christ, the channel was captured from the east. It shifted abruptly and flowed in that direction for about a thousand years. In the second century A.D., it was captured again, and taken south, by the now unprepossessing Bayou Lafourche, which, by the year 1000, was losing its hegemony to the river's present course, through the region that would be known as Plaquemines. By the nineteen-fifties, the Mississippi River had advanced so far past New Orleans and out into the Gulf that it was about to shift again, and its offspring Atchafalaya was ready to receive it. By the route of the Atchafalaya, the distance across the delta plain was a hundred and forty-five miles—well under half the length of the route of the master stream.

For the Mississippi to make such a change was completely natural, but in the interval since the last shift Europeans had settled beside the river, a nation had developed, and the nation could not afford nature. The consequences of the Atchafalaya's conquest of the Mississippi would include but not be limited to the demise of Baton Rouge and the virtual destruction of New Orleans. With its fresh water gone, its harbor a silt bar, its economy disconnected from inland commerce, New Orleans would turn into New Gomorrah. Moreover, there were so many big industries between the two cities that at night they made the river glow like a worm. As a result of settlement patterns, this reach of the Mississippi had long been known as "the German coast," and now, with B. F. Goodrich, E. I. du Pont, Union Carbide, Reynolds Metals, Shell, Mobil, Texaco, Exxon, Monsanto, Uniroyal, Georgia-Pacific, Hydrocarbon Industries, Vulcan Materials, Nalco Chemical, Freeport Chemical, Dow Chemical, Allied Chemical, Stauffer Chemical, Hooker Chemicals, Rubicon Chemicals, American Petrofina—with an infrastructural concentration equalled in few other places—it was often called "the American Ruhr." The industries were there because of the river. They had come for its navigational convenience and its fresh water. They would not, and could not, linger beside a tidal creek. For nature to take its course was simply unthinkable. The Sixth World War would do less damage to southern Louisiana. Nature, in this place, had become an enemy of the state.

Rabalais works for the U.S. Army Corps of Engineers. Some years ago, the Corps made a film that showed the navigation lock and a complex of associated structures built in an effort to prevent the capture of the Mississippi. The narrator said, "This nation has a large and powerful adversary. Our opponent could cause the United States to lose nearly all her seaborne commerce, to lose her standing as first among trading nations. . . . We are fighting Mother Nature. . . . It's a battle we have to fight day by day, year by year; the health of our economy depends on victory."

Rabalais was in on the action from the beginning, working as a construction inspector. Here by the site of the navigation lock was where the battle had begun. An old meander bend of the Mississippi was the conduit through which water had been escaping into the Atchafalaya. Complicating the scene, the old meander bend had also served as the mouth of the Red River. Coming in from the northwest, from Texas via Shreveport, the Red River had been a tributary of the Mississippi for a couple of thousand years—until the nineteen-forties, when the Atchafalaya captured it and drew it away. The capture of the Red increased the Atchafalaya's power as it cut down the country beside the Mississippi. On a map, these entangling watercourses had come to look like the letter "H." The Mississippi was the right-hand side. The Atchafalaya and the captured Red were the left-hand side. The crosspiece, scarcely seven miles long, was the former meander bend, which the people of the parish had long since named Old River. Sometimes enough water would pour out of the Mississippi and through Old River to quintuple the falls at Niagara. It was at Old River that the United States was going to lose its status among the world's trading nations. It was at Old River that New Orleans would be lost, Baton Rouge would be lost. At Old River, we would lose the American Ruhr. The Army's name for its operation there was Old River Control.

Rabalais gestured across the lock toward what seemed to be a pair of placid lakes separated by a trapezoidal earth dam a hundred feet high. It weighed five million tons, and it had stopped Old River. It had cut Old River in two. The severed ends were sitting there filling up with weeds. Where the Atchafalaya had entrapped the Mississippi, bigmouth bass were now in charge. The navigation lock had been dug beside this monument. The big dam, like the lock, was fitted into the mainline levee of the Mississippi. In Rabalais's pickup, we drove on the top of the dam, and drifted as well through Old River country. On this day, he said, the water on the Mississippi side was eighteen feet above sea level, while the water on the Atchafalaya side was five feet above sea level. Cattle were grazing on the slopes of the levees, and white horses with white colts, in deep-green grass. Behind the levees, the fields were flat and reached to rows of distant trees. Very early in the morning, a low fog had covered the fields. The sun, just above the horizon, was large and ruddy in the mist, rising slowly, like a hot-air balloon. This was a countryside of corn and soybeans, of grain-fed-catfish ponds, of feed stores and Kingdom Halls in crossroad towns. There were small neat cemeteries with ranks of white sarcophagi raised a foot or two aboveground, notwithstanding the protection of the levees. There were tarpapered cabins on concrete pylons, and low brick houses under planted pines. Pickups under the pines. If this was a form of battlefield, it was not unlike a great many battlefields—landscapes so quiet they belie their story. Most battlefields, though, are places where something happened once. Here it would happen indefinitely.

We went out to the Mississippi. Still indistinct in mist, it looked like a piece of the sea. Rabalais said, "That's a wide booger, right there." In the spring high water of vintage years—1927, 1937, 1973—more than two million cubic feet of water had gone by this place in every second. Sixty-five kilotons per second. By the mouth of the inflow channel leading to the lock were rock jetties, articulated concrete mattress revetments, and other heavy defenses. Rabalais observed that this particular site was no more vulnerable than almost any other point in this reach of river that ran so close to the Atchafalaya plain. There were countless places where a break-out might occur: "It has a tendency to go through just anywheres you can call for."

Why, then, had the Mississippi not jumped the bank and long since diverted to the Atchafalaya?

"Because they're watching it close," said Rabalais. "It's under close surveillance." ■

Rachel Carson

FROM "The Marginal World" in *The Edge of the Sea* (1955)

The edge of the sea is a strange and beautiful place. All through the long history of Earth it has been an area of unrest where waves have broken heavily against the land, where the tides have pressed forward over the continents, receded, and then returned. For no two successive days is the shore line precisely the same. Not only do the tides advance and retreat in their eternal rhythms, but the level of the sea itself is never at rest. It rises or falls as the glaciers melt or grow, as the floor of the deep ocean basins shifts under its increasing load of sediments, or as the earth's crust along the continental margins warps up or down in adjustment to strain and tension. Today a little more land may belong to the sea, tomorrow a little less. Always the edge of the sea remains an elusive and indefinable boundary.

The shore has a dual nature, changing with the swing of the tides, belonging now to the land, now to the sea. On the ebb tide it knows the harsh extremes of the land world, being exposed to heat and cold, to wind, to rain and drying sun. On the flood tide it is a water world, returning briefly to the relative stability of the open sea.

Only the most hardy and adaptable can survive in a region so mutable, yet the area between the tide lines is crowded with plants and animals. In this difficult world of the shore, life displays its enormous toughness and vitality by occupying almost every conceivable niche. Visibly, it carpets the intertidal rocks; or half hidden, it descends into fissures and crevices, or hides under boulders, or lurks in the wet gloom of sea caves. Invisibly, where the casual observer would say there is no life, it lies deep in the sand, in burrows and tubes and passageways. It tunnels into solid rock and bores into peat and clay. It encrusts weeds or drifting spars or the hard, chitinous shell of a lobster. It exists minutely, as the film of bacteria that spreads over a rock surface or a wharf piling; as spheres of protozoa, small as pinpricks, sparkling at the surface of the sea; and as Lilliputian beings swimming through dark pools that lie between the grains of sand.

The shore is an ancient world, for as long as there has been an earth and sea there has been this place of the meeting of land and water. Yet it is a world that keeps alive the sense of continuing creation and of the relentless drive of life. Each time that I enter it, I gain some new awareness of its beauty and its deeper meanings, sensing that intricate fabric of life by which one creature is linked with another, and each with its surroundings.

In my thoughts of the shore, one place stands apart for its revelation of exquisite beauty. It is a pool hidden within a cave that one can visit only rarely and briefly when the lowest of the year's low tides fall below it, and perhaps from that very fact it acquires some of its special beauty. Choosing such a tide, I hoped for a glimpse of the pool. The ebb was to fall early in the morning. I knew that if the wind held from the northwest and no interfering swell ran in from a distant storm the level of the sea should drop below the entrance to the pool. There had been sudden ominous showers in the night, with rain like handfuls of gravel flung on the roof. When I looked out into the early morning the sky was full of a gray dawn light but the sun had not yet risen. Water and air were pallid. Across the bay the moon was a luminous disc in the western sky, suspended above the dim line of distant shore—the full August moon, drawing the tide to the low, low levels of the threshold of the alien sea world. As I watched, a gull flew by, above the spruces. Its breast was rosy with the light of the unrisen sun. The day was after all to be fair.

Later, as I stood above the tide near the entrance to the pool, the promise of that rosy light was sustained. From the base of the steep wall of rock on which I stood, a moss-covered ledge jutted seaward into deep water. In the surge at the rim of the ledge the dark fronds of oarweeds swayed, smooth and gleaming as leather. The projecting ledge was the path to the small hidden cave and its pool. Occasionally a swell, stronger than the rest, rolled smoothly over the rim and broke in foam against the cliff. But the intervals between such swells were long enough to admit me to the ledge and long enough for a glimpse of that fairy pool, so seldom and so briefly exposed.

And so I knelt on the wet carpet of sea moss and looked back into the dark cavern that held the pool in a shallow basin. The floor of the cave was only a few inches below the roof, and a mirror had been created in which all that grew on the ceiling was reflected in the still water below.

Under water that was clear as glass the pool was carpeted with green sponge. Gray patches of sea squirts glistened on the ceiling and colonies of soft coral were a pale apricot color. In the moment when I looked into the cave a little elfin starfish hung down, suspended by the merest thread, perhaps by only a single tube foot. It reached down to touch its own reflection, so perfectly delineated that there might have been, not one starfish, but two. The beauty of the reflected images and of the limpid pool itself was the poignant beauty of things that are ephemeral, existing only until the sea should return to fill the little cave.

Whenever I go down into this magical zone of the low water of the spring tides, I look for the most delicately beautiful of all the shore's inhabitants—flowers that are not plant but animal, blooming on the threshold of the deeper sea. In that fairy cave I was not disappointed. Hanging from its roof were the pendent flowers of the hydroid Tubularia, pale pink, fringed and delicate as the wind flower. Here were creatures so exquisitely fashioned that they seemed unreal, their beauty too fragile to exist in a world of crushing force. Yet every detail was functionally useful, every stalk and hydranth and petal-like tentacle fashioned for dealing with the realities of existence. I knew that they were merely waiting, in that moment of the tide's ebbing, for the return of the sea. Then in the rush of water, in the surge of surf and the pressure of the incoming tide, the delicate flower heads would stir with life. They would sway on their slender stalks, and their long tentacles would sweep the returning water, finding in it all that they needed for life.

And so in that enchanted place on the threshold of the sea the realities that possessed my mind were far from those of the land world I had left an hour before. In a different way the same sense of remoteness and of a world apart came to me in a twilight hour on a great beach on the coast of Georgia. I had come down after sunset and walked far out over sands that lay wet and gleaming, to the very edge of the retreating sea. Looking back across that immense flat, crossed by winding, water-filled gullies and here and there holding shallow pools left by the tide, I was filled with awareness that this intertidal area, although abandoned briefly and rhythmically by the sea, is always reclaimed by the rising tide. There at the edge of low water the beach with its reminders of the land seemed far away. The only sounds were those of the wind and the sea and the birds. There was one sound of wind moving over water, and another of water sliding over the sand and tumbling down the faces of its own wave forms. The flats were astir with birds, and the voice of the willet rang insistently. One of them stood at the edge of the water and gave its loud, urgent cry; an answer came from far up the beach and the two birds flew to join each other. ▪

7 The Work of Ice

In the mountains it's cold.
Always been cold, not just this year.
Jagged scarps forever snowed in
Woods in the dark ravines spitting mist.
Grass is still sprouting at the end of June,
Leaves begin to fall in early August.
And here am I, high on mountains,
Peering and peering, but I can't even see the sky.

Men ask the way to Cold Mountain
Cold Mountain: there's no through trail.
In summer, ice doesn't melt
The rising sun blurs in swirling fog.
How did I make it?
My heart's not the same as yours.
If your heart was like mine
You'd get it and be right here.

> —Han-Shan, "Cold Mountain Poems"
> *translated by Gary Snyder*

IMAGINE A COLD WORLD. IMAGINE GLACIERS AND ICE SHEETS, MASSES OF ice formed by recrystallized snow, flowing under their own weight. Great ice sheets covered large parts of Earth, from the poles to regions now temperate, at many different times in the past. Geoscientists believe that the most recent major ice ages occurred in the Pleistocene Epoch, between two million and ten thousand years ago, during which several major advances and retreats of glaciers occurred in response to repeated warming and cooling of Earth's climate and oceans. During the most recent ice age, estimated at seventy-five to ten thousand years ago, over one-quarter of Earth's land area, including almost all of Canada and the northern United States, lay under ice.

The last major continental ice sheets melted back approximately fifteen thousand years ago, leaving only a few mountain glaciers, the Greenland ice sheet, and the great Antarctic ice sheet, the latter about the size of the lower forty-eight United States. Ice now covers one-tenth of the world's land surface. Despite the meltback of the ice sheets, still almost all the fresh water on Earth is contained in ice.

Like flowing water and wind, glaciers work as agents of erosion and deposition, but they are much more effective at eroding Earth's surface. By plucking fragments from solid rock, by abrading landscapes, glacier movement creates distinctive topography and landforms, from striations, grooves, and polish on bedrock surfaces to erosional forms left by valley glaciers in mountainous areas including cirques, horns, aretes, cols, U-shaped valleys, hanging valleys, and fjords. The eroded debris transported by glaciers, from boulders to fine rock flour, are eventually deposited as moraines where the ice ends or as outwash by meltwater streams. Debris from the last major continental ice sheets form such well-known geographic features as Long Island, New York, and Cape Cod, Massachusetts.

The selections in this chapter considers the work of flowing ice, melted ice, and wind blown beyond the ice edge. We begin with the prologue of John Imbrie and Katherine Palmer Imbrie's *Ice Ages: Solving the Mystery* (1979, 1986), which presents a historical overview of the most recent ice age and the beginnings of climate science.

William Scoresby, Jr. (1789–1857), son of a whale fishery magnate, was an

English Arctic explorer, scientist, and cleric. After his first voyage at the age of eleven, he served as chief officer of the whaler *Resolution* in 1806 when that ship reached 81° 30' N. latitude, for more than twenty years the highest known northern latitude sailed to in the Eastern Hemisphere. Scoresby wrote many scientific papers and books about the Arctic—and spent nearly two-score summers sailing in polar waters—with emphases on magnetism and the ship's compass, ice formation on land and sea, submarine life, structures of snowflakes, and characteristics of ocean water. An excerpt from *An Account of the Arctic Regions, with a History and Description of the Northern Whale-Fishery* (1820) presents his astute observations of ice.

Contemporary writer and historian Stephen J. Pyne writes about the Earth's southern polar region in *The Ice: A Journey to Antarctica*. Writing more than a century and a half after Scoresby, Pyne reflects with similar enthusiasm and detail about "a world informed by ice."

In "The Great Scablands Debate" Stephen Jay Gould (1941–2002), prominent evolutionary biologist and author, examines the changing interpretations of the origin of the channeled scablands in eastern Washington State, features now considered by geologists to have formed by a tremendous flood of glacial meltwater at the end of the last ice age. A key message conveyed in the piece is that "dogmas play their worst role when they lead scientists to reject beforehand a counterclaim that could be tested in nature."

Swiss geologist and alpinist August Gansser (1910–) played an eminent role in determining the geological structures of the Himalaya. From his book with Arnold Heim *The Throne of the Gods* (1939) we include his account of a harrowing 1936 crossing, with three companions, of the ice- and snow-covered, 18,400-foot Traill Pass that lies between Nanda Devi and Nanda Kot in the central Himalaya.

In a selection from the novel *Wolfsong* (1991), Choctaw-Cherokee-Irish author and scholar Louis Owens (1948–2002) describes an eroded and filled landscape of the glaciated North Cascades of Washington.

We end the section by returning to John Muir (1838–1914) and his experiences in California's Sierra Nevada. In the excerpt from *The Yosemite* (1912), Muir describes with clear understanding the glacial history of Yosemite Valley.

John Imbrie and Katherine Palmer Imbrie

FROM *Ice Ages: Solving the Mystery* (1979, 1986)

Prologue: The Forgotten Ice Age

Twenty-thousand years ago, the earth was held in thrall by relentlessly probing fingers of ice—ice that drew its power from frigid strongholds in the north, and flowed southward to bury forests, fields, and mountains. Landscapes that were violated by the slowly moving glaciers would carry the scars of this advance far into the future. Temperatures plummeted, and land surfaces in many parts of the world were depressed by the unrelenting weight of the thrusting ice. At the same time, so much water was drawn from the ocean to form these gargantuan glaciers that sea levels around the world fell by 350 feet, and large areas of the continental shelf became dry land.

This period in the earth's history has come to be called the ice age. In North America, glacial ice spread out from centers near Hudson Bay to bury all of eastern Canada, New England, and much of the Midwest under a sheet of ice that averaged more than a mile in thickness. A second ice sheet spread out from centers in the Canadian Rockies and other highlands in western North America to engulf parts of Alaska, all of western Canada, and portions of Washington, Idaho, and Montana. In Europe, the ice reached outward from Scandinavia and Scotland to cover most of Great Britain, Denmark, and large parts of northern Germany, Poland, and the Soviet Union. A smaller ice cap, centered on the Alps, buried all of Switzerland and nearby portions of Austria, Italy, France, and Germany. In the southern hemisphere, small ice sheets developed over parts of Australia, New Zealand, and Argentina. In all, the ice covered about 11 million square miles of land that is today free of ice.

Immediately south of these great northern-hemisphere ice sheets, the landscape was treeless tundra. Here, during the short, cool summers, heather and other hardy, low-growing plants grew in the boggy soil. Migrating herds of reindeer and mammoths grazed upon this lush plant cover during the summer months, and in winter moved southward seeking more favorable pastures. In North America, the tundra was only a narrow belt of land that served to separate the ice sheets to the north from forested areas to the south. In the eastern part of the continent, spruce trees grew in a continuous forest; in the more arid Midwest, the stands of spruce followed the rivers, while in between there were dusty grasslands.

In Europe and Asia, the tundra belt was wider than it was in North America, giving way only gradually to a vast expanse of semiarid grassland that stretched from horizon to horizon—across two continents from the Atlantic coast of France, through central Europe, to eastern Siberia.

Stone Age hunters, following herds of mammoth and reindeer across the tundra, could glimpse the southern edge of the ice sheet. As the cold penetrated their deerhide clothing, and the wind from the north whipped their faces, it would have been difficult for these people to realize that their descendants would inhabit a very different world from their own.

Yet the ice age did come to an end. About 14,000 years ago, the ice sheets began to retreat. Within 7,000 years they had withdrawn to their present limits. Today, all that remains of the ice sheets in the northern hemisphere are the Greenland Ice Sheet and a few small ice caps in the Canadian Arctic. Where modern farmers reap Iowa corn and Dakota wheat, mile-high glaciers once ground their way over the land. And, where European forests stand today, treeless plains once stretched to the horizon.

As the glaciers melted back, the landscape they left behind was greatly altered—a landscape strewn with traces of its glacial origin. In northern regions, the ice sheets had ground away at the underlying surface, scratching deep grooves in the bedrock, and swallowing bits and pieces of eroded material. This material had been transported outward to the margins of the ice sheets where it was deposited in a chaotic jumble known as a moraine.

As the ice sheets withdrew, human recollections of them began to fade. Racial memory—if it exists at all—must be imperfect, for the world of the Stone Age hunters was soon forgotten. Even the clues left behind by the great ice sheets were misinterpreted. By the eighteenth century, geologists surmised that the blanket of glacial sediments had been transported and deposited by the great flood described in the Bible. It was only in the early years of the nineteenth century that some scientists began to question this explanation. Were floodwaters—even divinely inspired ones—actually capable of transporting gigantic boulders hundreds of miles, or was some other agent responsible? ▪

William Scoresby, Jr.

FROM *An Account of the Arctic Regions, with a History and Description of the Northern Whale-Fishery* (1820)

Of the inanimate productions of the Polar Seas, none perhaps excites so much interest and astonishment in a stranger, as the *ice* in its great abundance and variety. The stupendous masses, known by the name of *Ice-islands*, or *Ice-bergs*, common to Davis' Straits and sometimes met with in the Spitzbergen Sea, from their height, various forms, and the depth of water in which they ground, are calculated to strike the beholder with wonder; yet the prodigious sheets of ice, called *fields*, more peculiar to the Spitzbergen Sea, are not less astonishing. Their deficiency in elevation, is sufficiently compensated by their amazing extent of surface. Some of them have been observed extending many leagues in length, and covering an area of several hundreds of square miles; each consisting of a single sheet of ice, having its surface raised in general four or six feet above the level of the water, and its base depressed to the depth of ten to twenty feet beneath.

The ice in general is designated by a variety of appellations, distinguishing it according to the size or shape of the pieces, their number or form of aggregation, thickness, transparency, situation, &c.

As the different denominations of ice will be frequently referred to in the course of this work, it may be useful to give definitions of the terms in use among the whale-fishers, for distinguishing them.

1. An *ice-berg* or ice mountain, is a large insulated peak of floating ice; or a glacier, occupying a ravine or valley, generally opening towards the sea, in an arctic country.

2. A *field* is a sheet of ice so extensive, that its limits cannot be discerned from a ship's mast-head.

3. A *floe* is similar to a field, but smaller; inasmuch as its extent *can* be seen. This term, however, is seldom applied to pieces of ice of less diameter than half a mile or a mile.

4. *Drift-ice* consists of pieces less than floes, of various shapes and magnitudes.

5. *Brash-ice* is still smaller than drift-ice, consisting of roundish nod-
   ules, and fragments of ice, broken off by the attrition of one piece
   against another. This may be considered as the wreck of other kinds
   of ice.

6. *Bay-ice* is that which is newly formed on the sea, and consists of
   two kinds, common bay-ice, and *pancake-ice*; the former occurring
   in smooth extensive sheets, and the latter in small circular pieces
   with raised edges.

7. *Sludge* consists of a stratum of detached ice-crystals, or of snow, or
   of the smaller fragments of brash-ice floating on the surface of the
   sea. This generally forms the rudiments of ice, when the sea is in
   agitation.

[. . .]

Ice-bergs differ a little in color, according to their solidity and distance, or
state of the atmosphere. A very general appearance is that of cliffs of chalk,
or of white or grey marble. The sun's rays reflected from them, sometimes
give a glistening appearance to their surfaces. Different shades of color
occur in the precipitous parts, accordingly as the ice is more or less solid,
and accordingly as it contains strata of earth, gravel, or sand, or is free from
any impurity. In the fresh fracture, greenish-grey, approaching to emerald-
green, is the prevailing color.

In the night, ice-bergs are readily distinguished, even at a distance, by
their natural effulgence; and, in foggy weather, by a peculiar blackness in
the atmosphere, by which the danger to the navigator is diminished. As,
however, they occur far from land, and often in unexpected situations, nav-
igators crossing the Atlantic in the gloom of night, between the parallels
of 50° and 60° of latitude, or even farther to the south, require to be always
on the watch for them. In some places, near Cape Farewell, or towards the
mouth of Davis' Strait, they sometimes occur in extensive chains; in
which case, fatal accidents have occurred, by vessels getting involved
among them in the night, during storms. But ice-bergs occurring singly,
have rarely been productive of any serious mischief.

[. . .]

It is common, when ships moor to icebergs, to lie as remote from them as their ropes will allow, and yet accidents sometimes happen, though the ship ride at the distance of a hundred yards from the ice. Thus, *calves* rising up with a velocity nearly equal to that of the descent of a falling berg, have produced destructive effects. In the year 1812, while the Thomas of Hull, Captain Taylor, lay moored to an ice-berg in Davis' Strait, a *calf* was detached from beneath, and rose with such tremendous force, that the keel of the ship was lifted on a level with the water at the bow, and the stern was nearly immersed beneath the surface. Fortunately the blow was received on the keel, and the ship was not materially damaged.

From the deep pools of water formed in the summer season, on the depressed surface of some bergs, or from the streams running down their sides, the ships navigating where they abound, are presented with opportunities for watering with the greatest ease and dispatch. For this purpose, casks are landed upon the lower bergs, filled, and rolled into the sea; but from the higher, the water is conveyed by means of a long tube of canvas or leather, called a *hose,* into casks placed in the boats, at the side of the ice, or even upon the deck of a ship.

The greater part of the ice-bergs that occur in Davis' Strait, and on the eastern coast of North America, notwithstanding their profusion and immense magnitude, seem to be merely fragments of the land ice-bergs or glaciers, which exist in great numbers on the coast, forming the boundaries of Baffin's Bay. These glaciers fill immense valleys, and extend in some places several miles into the sea; in others, they terminate with a precipitous edge at the general line formed by the coast. In the summer season, when they are particularly fragile, the force of cohesion is often overcome by the weight of the prodigious masses that overhang the sea; and in winter, the same effect may be produced, by the powerful expansion of the water filling any excavation or deep-seated cavity, when its dimensions are enlarged by freezing, thereby exerting a tremendous force, and bursting the berg asunder. ∎

Stephen J. Pyne

FROM *The Ice: A Journey to Antarctica* (1986)

To enter Greater Antarctica is to be drawn into a slow maelstrom of ice. Ice is the beginning of Antarctica and ice is its end. As one moves from perimeter to interior, the proportion of ice relentlessly increases. Ice creates more ice, and ice defines ice. Everything else is suppressed. This is a world derived from a single substance, water, in a single crystalline state, snow, transformed into a lithosphere composed of a single mineral, ice. This is earthscape transfigured into icescape. Here is a world informed by ice: ice that welds together a continent: ice on such a scale that it shapes and defines itself: ice that is both substance and style: ice that is both landscape and allegory. The berg is a microcosm of this world. It is the first and, paradoxically, the most complex materialization of The Ice. It is a fragment torn loose from the bottom of the globe, the icy underworld of the Earth; from the ends of the world, its past and future; from the Earth's polar source, the end that makes possible the means. The berg is both substance and symbol. "Everything is in it," as Conrad wrote of the human mind, "all the past as well as all the future." The journey of the ice from core to margin, from polar plateau to open sea, narrates an allegory of mind and matter.

The great berg spins in a slow, counterclockwise gyre.

It is only another of a series of rotations that have characterized the berg's fantastic journey. The continental plates that comprise the land form a lithospheric mosaic and spin with the infinitesimal patience of geologic time; the Southern Ocean courses around them, the gyre of the circumpolar current; storm cells swirl over the ocean, epicycles of the polar vortex; sea ice floes, like a belt of asteroids, circle endlessly, a life cycle of freezing and melting; icebergs, large and small, circle like comets around their peculiar icy sun. Superimposed over all these motions, the Earth itself rotates around its pole and revolves around the Sun. The ice terranes ring the core like concentric crystalline spheres. The ice mass that became the berg has passed from ice dome to sheet ice to glacier ice to shelf ice to pack ice to the diminutions of the bergs, cycle by cycle, like the gears of an ice orrery. The large bergs fragment into smaller bergs, the small bergs into bergy bits, the bits into growlers, the growlers into brash ice, the brash into chips and meltwater. With each outward frontier the pace of activity quickens.

Ice informs the geophysics and geography of Antarctica. It connects land to land, land to sea, sea to air, air to land, ice to ice. The Antarctic atmosphere consists of ice clouds and ice vapor. The hydrosphere exists as ice rivers and ice seas. The lithosphere is composed of ice plateaus and ice mountains. Even those features not completely saturated with ice are vastly reduced. The atmosphere is much thinner at the poles than elsewhere, in part because of the great height of the polar ice sheet. The hydrosphere is charged with bergs and coated with ice floes; during the polar night, its cover of sea ice effectively doubles the total ice field of Antarctica. The lithosphere is little more than a matrix for ice. Less than 3 percent of Antarctica consists of exposed rock, and the rock is profoundly influenced by periglacial processes, an indirect manifestation of ice.

Out of simple ice crystals is constructed a vast hierarchy of ice masses, ice terranes, and ice structures. These higher-order ice forms collectively compose the entire continent: *the ice-bergs:* tabular bergs, glacier bergs, ice islands, bergy bits, growlers, brash ice, white ice, blue ice, green ice, dirty ice; *the sea ices:* pack ice, ice floes, ice rinds, ice hummocks, ice ridges, ice flowers, ice stalactites, pancake ice, frazil ice, grease ice, congelation ice, infiltration ice, undersea ice, vuggy ice, new ice, old ice, brown ice, rotten ice; *the coastal ices:* fast ice, shore ice, glacial-ice tongues, ice piedmonts, ice fringes, ice cakes, ice foots, ice fronts, ice walls, floating ice, grounded ice, anchor ice, rime ice, ice ports, ice shelves, ice rises, ice bastions, ice haycocks, ice lobes, ice streams; *the mountain ices:* glacial ice, valley glaciers, cirque glaciers, piedmont glaciers, ice fjords, ice layers, ice pipes, ice falls, ice folds, ice faults, ice pinnacles, ice lenses, ice aprons, ice falls, ice fronts, ice slush; *the ground ices:* ice wedges, ice veins, permafrost; *the polar plateau ices:* ice sheets, ice caps, ice domes, ice streams, ice divides, ice saddles, ice rumples; *the atmospheric ices:* ice grains, ice crystals, ice dust, pencil ice, plate ice, bullet ice. The ice field is organized into a series of roughly concentric ice terranes, like the ordered rings comprising the hierarchy of Dante's cosmology.

[. . .]

[. . .] The Earth's cryosphere joins it to other worlds and other times—to the outer solar system and to vanished geologic pasts. It is a white warp in space and time. That the Earth's ice consists not of ammonia or carbon dioxide or methane but of water, that it is crystalline rather than amorphous, and that the planet's temperature range falls within the triple point of water account for the Earth's uniqueness, and dynamism, as a member of the ice cosmos.

The berg contains a record of all this. Its travels have a mythic quality, a retrograde journey out of an underworld. It is a voyage that joins microcosm to macrocosm, that builds from a single substance—ice crystals—a vast, almost unbounded continent. Yet a descent to this underworld—from the ice-induced fog that shrouds the continent to the unblinking emptiness that commands its center—does not lead to more splendid scenes, as a trip through the gorge of the Grand Canyon does, or to richer displays of life, as a voyage to the interior of the Amazon does, or to more opulent civilizations, living or dead, as the excavation of Egypt's Valley of the Kings or the cities of Troy does, or to greater knowledge, as the ultimately moral journeys of Odysseus, Aeneas, Dante, even Marlowe do. It leads only to more ice. Almost everything is there because almost nothing is there.

Antarctica is the Earth's great sink, not only for water and heat but for information. Between core and margin there exist powerful gradients of energy and information. These gradients measure the alienness of The Ice as a geographic and cultural entity. The Ice is profoundly passive: it does not give, it takes. The Ice is a study in reductionism. Toward the interior everything is simplified. The Ice absorbs and, an imperfect mirror, its ineffable whiteness reflects back what remains. Toward the perimeter, ice becomes more complex, its shapes multiply, and its motions accelerate. The ephemeral sample, the berg, is more interesting than the invariant whole, the plateau. The extraordinary isolation of Antarctica is not merely geophysical but metaphysical. Cultural understanding and assimilation demand more than the power to overcome the energy gradient that surrounds The Ice: they demand the capacity and desire to overcome the information gradient. Of all the ice masses in Greater Antarctica the berg is the most varied, the most informative, and the most accessible. The assimilation of The Ice begins with the assimilation of the iceberg. ■

Stephen Jay Gould

FROM "The Great Scablands Debate" in *The Panda's Thumb:
More Reflections in Natural History* (1980)

In this essay I tell a local, geologic story. But it conveys the same message—that dogmas play their worst role when they lead scientists to reject beforehand a counterclaim that could be tested in nature.

Flow basalts of volcanic origin blanket most of eastern Washington. These basalts are often covered by a thick layer of loess, a fine-grained, loosely packed sediment blown in by winds during the ice ages. In the area between Spokane and the Snake and Columbia rivers to the south and west, many spectacular, elongate, subparallel channel-ways are gouged through the loess and deeply into the hard basalt itself. These coulees, to use the local name, must have been conduits for glacial meltwaters, for they run down gradient from an area near the southern extent of the last glacier into the two major rivers of eastern Washington. The channeled scablands—as geologists designate the entire area—are puzzling as well as awesome, and for several reasons:

1. The channels connect across tall divides that once separated them. Since the channels are hundreds of feet deep, this extensive anastomosis indicates that a prodigious amount of water must once have flowed over the divide.

2. As another item favoring channels filled to the brim with water, the sides of the coulees contain many hanging valleys where tributaries enter the main channels. (A hanging valley is a tributary channel that enters a main channel high above the main channel's modern stream bed.)

3. The hard basalt of the coulees is deeply gouged and scoured. This pattern of erosion does not look like the work of gentle rivers in the gradualist mode.

4. The coulees often contain a number of high-standing hills composed of loess that has not been stripped away. These are arranged as if they were once islands in a gigantic braided stream.

5. The coulees contain discontinuous deposits of basaltic stream gravel, often composed of rock foreign to the local area.

Just after World War I, Chicago geologist J Harlen Bretz advanced an unorthodox hypothesis to account for this unusual topography (yes, that's J without a period, and don't ever let one slip in, for his wrath can be terrible). He argued that the channeled scablands had been formed all at once by a single, gigantic flood of glacial meltwater. This local catastrophe filled the coulees, cut through hundreds of feet of loess and basalt, and then receded in a matter of days. He ended his major work of 1923 with these words:

> Fully 3,000 square miles of the Columbia Plateau were swept by the glacial flood, and the loess and silt cover removed. More than 2,000 square miles of this area were left as bare, eroded rock-cut channel floors, now the scablands, and nearly 1,000 square miles carry gravel deposits derived from the eroded basalt. It *was* a debacle which swept the Columbia Plateau.

Bretz's hypothesis became a minor *cause célèbre* within geological circles. Bretz's stout and lonely defense of his catastrophic hypothesis won some grudging admiration, but virtually no support at first. The "establishment," as represented by the United States Geological Survey, closed ranks in opposition. They had nothing better to propose, and they did admit the peculiar character of scabland topography. But they held firm to the dogma that catastrophic causes must never be invoked so long as any gradualist alternative existed. Instead of testing Bretz's flood on its own merits, they rejected it on general principles.

On January 12, 1927, Bretz bearded the lion in its lair and presented his views at the Cosmos Club, in Washington, D.C., before an assembled group of scientists, many from the Geological Survey. The published discussion clearly indicates that a priori gradualism formed the basis for Bretz's glacial reception. I include typical comments from all detractors.

W. C. Alden admitted "it is not easy for one, like myself, who has never examined this plateau to supply offhand an alternative explanation of the phenomena." Nonetheless, undaunted, he continued: "The main difficulties seem to be: (1) The idea that all the channels must have been developed simultaneously in a very short time; and (2) the tremendous amount of water that he postulates. . . . The problem would be easier if less water was required and if longer time and repeated floods could be allotted to do the work."

James Gilluly, this century's chief apostle of geological gradualism, ended a long comment by noting "that the actual floods involved at any given time were of the order of magnitude of the present Columbia's or at most a few times as large, seems by no means excluded by any evidence as yet presented."

E. T. McKnight offered a gradualist alternative for the gravels: "This writer believes them to be the normal channel deposits of the Columbia during its eastward shift over the area in preglacial, glacial, and postglacial times."

G. R. Mansfield doubted that "so much work could be done on basalt in so short a time." He also proposed a calmer explanation: "The scablands seem to me better explained as the effects of persistent ponding and overflow of marginal glacial waters, which changed their position or their places of outlet from time to time through a somewhat protracted period."

Finally, O. E. Meinzer admitted that "the erosion features of the region are so large and bizarre that they defy description." They did not, however, defy gradualist explanation: "I believe the existing features can be explained by assuming normal stream work of the ancient Columbia River." Then, more baldly than most of his colleagues, he proclaimed his faith: "Before a theory that requires a seemingly impossible quantity of water is fully accepted, every effort should be made to account for the existing features without employing so violent an assumption."

The story has a happy ending, at least from my point of view, for Bretz was delivered from the lion's lair by later evidence. Bretz's hypothesis has prevailed, and virtually all geologists now believe that catastrophic floods cut the channeled scablands. Bretz had found no adequate source for his floodwaters. He knew that the glaciers had advanced as far as Spokane, but neither he nor anyone else could imagine a reasonable way to melt so much water so rapidly. Indeed, we still have no mechanism for such an episodic melting.

The solution came from another direction. Geologists found evidence for an enormous, ice-dammed glacial lake in western Montana. This lake emptied catastrophically when the glacier retreated and the dam broke. The spillway for its waters leads right into the channeled scablands.

Bretz had presented no really direct evidence for deep, surging water. Gouging might have proceeded sequentially, rather than all at once; anastomosis and hanging valleys might reflect filled coulees with gentle, rather than raging, flow. But when the first good aerial photographs of the scablands were taken, geologists noticed several areas on the coulee floors are covered with giant stream bed ripples, up to 22 feet high and 425 feet long. Bretz, like an ant on a Yale bladderball, had been working on the wrong scale. He had been walking over the ripples for decades but had been too close to see them. They are, he wrote quite correctly, "difficult to identify at ground level under a cover of sagebrush." Observations can only be made at appropriate scales.

Hydraulic engineers can infer the character of flow from the size and shape of ripples on the stream bed. V. R. Baker estimates a maximum discharge of 752,000 cubic feet per second in the scabland flow channels. Such a flood could have moved 36-foot boulders. ∎

August Gansser

FROM Arnold Heim and August Gansser's
The Throne of the Gods (1939)

While Dewan Singh was getting his own equipment together and providing himself with food for several days, we started ahead up the savage Lwaln Valley, along a narrow footpath which soon lost itself in dense rhododendron thickets. Beyond the head of the valley two mighty cones beckoned, those of Nanda Devi, 25,584, and 24,272 feet. Gradually the higher western peak was eclipsed by the eastern one. Dewan Singh was not yet in sight. We intended to reach the foot of the glacier that day. It was rather an ambitious plan, but we carried it out at the cost of a good deal of sweating. We camped among alpine roses at 13,450 feet, just short of the glacier. Soon Dewan Singh turned up and sat down with us beside our blazing camp fire. We did not bother to pitch our little tent, but slept in the open.

"August 26th. Yesterday's sunset, of a rich and rare violet tint like a picture postcard, is followed this morning, between clouds, by a fiery red sunrise. Still we do not long enjoy this rich play of color for a monotonous grey soon succeeds. We cross the unnamed glacier which flows down from the Nanda Kot side—that is to say, we cross the endless moraine, seeing very little trace of ice. Like a river in flood, full to the brim, the glacier flows down the valley between the two lateral moraines. Thick clouds enfold the gap we wish to cross. To the west we can still see the 10,000 feet rampart of Nanda Devi. Quickly I make a geological sketch, the mist gathering as I do so.

"To the east I can dimly make out an ice-encrusted wall, terminating in a spade-shaped peak. It must be Nanda Kot. In good weather, this would be a splendid position, flanked as we are by the two Nandas. We decide to stay where we are and hope for better weather, but to-morrow we positively must get over the pass. I study the eastern face of Nanda Devi. As far as I can make out, there is very little metamorphic rock, and certainly no gneiss or granite as was formerly supposed. After this I sit in the little tent, which we have pitched for shelter, and write to the accompaniment of the too familiar melody of the raindrops which fall steadily, though not as yet very heavily. But to-morrow, what will the weather be like to-morrow? Enough to reduce a man to despair, this weather, to make him pluck out his beard by the roots. Ice and snow avalanches thunder down from Nanda Devi. Nanda Kot is better behaved, almost sanctimonious.

"The baked rice given us by the patwari of Milam is delicious, and the sweet mess of tsamba and brown sugar which he recommended me as a fortifier on this toilsome journey is excellent in tea. But this idling about plays havoc with our scanty rations. Now Nanda Kot is raging in his turn. He seems to wake up and growl when it snows on to his head and rains on to his belly.

"Night fell long since. I am lying in the tent and look at Nanda Kot which emerges from the clouds in the light of the half-moon. When I stretch forth my head and look in the opposite direction I see Nanda Devi like a ghost in the moonlight. Stars sparkle above it.

"August 27th. Stars sparkle above it, and the Milky Way shows in glory above the peak, as if it were smoking. The moon has set, and it is now the darkest hour before the dawn. Gradually a glimmer of light comes. The two Nandas shine with a silver light. Then Nanda Devi begins to redden. A glorious day is dawning. When we reach 15,400 feet, we come upon new-fallen snow. Hard frozen, it crackles beneath our feet as we trudge round the numerous crevasses. The large ones are partially filled with snow. We have to let ourselves down over an ice face, cautiously make our way over the bottom, and then scale the other face. Here, in most cases, there was a snow overhang through which we had to burrow. Having crossed the glacier, we began the ascent of the snow-slopes, high above which was an overhanging ice-crest, our main problem. We have plenty of time to study its flutings as we slowly and laboriously, cutting footholds, climb towards it. A great portion of the overhang breaks away, rushing towards us as an avalanche, but breaking up in the various runnels and racing safely past us. Nevertheless the danger increases, for avalanches of new-fallen snow are rushing down everywhere, and both the Nandas are thundering. When we look down, we see that our climb has become almost vertical. A new fragment of the overhang breaks away. This time it sweeps down as one mass, but happily misses us, though we feel the wind of its passing. The overhang still looms threateningly above, and the glacier, with its marginal gulf, is far, far below. The prospects are by no means rosy. Perhaps it is better not to look up. There may not be another avalanche before we get over. My companions steady their footing by thrusting their ice-axes home as they cautiously move from one to another along the footholds I have cut. I try to get round the overhang by moving a few paces to the left. Straight above me for 10 feet it is absolutely impassable. Very cautiously I use my ice-axe. The snow falls in heaps upon my sun-burned face. However cautiously, I must work furiously, for a minute's delay may lead to a general disaster. Yet there is no way back, now; the only chance of escape is to get over. Heavy masses of snow fall, jarring the rope, and almost wrenching me from my foothold. At length I can reach over the top with my ice-axe, thrust it deep into the snow and pull myself over. A moment or two, to take breath. Such strenuous exertion at 18,350 feet is very trying to the wind, but the rope is made fast. The others are pulled up safely, and when all are over the top we can relax a little from our 'do or die' mood."

The other side is pretty steep too, but not for long. Soon we get down to the glacier which forms the source of the Pindari.

"Traill Pass proper is only at 17,700 feet—650 feet lower than the altitude at which we crossed. It was owing to the recent heavy snowfalls and the amount of snow that lay on the rocks that we had to venture upon the steep ice of the declivity we actually scaled. Roped in pairs we go down the broad glacier. It has become intolerably hot. Often we sink in the snow up to our hips. Evening comes on. For a change, Kirken falls through the snow into a crevasse, but we are soon able to pull him out. By the last glimmer of daylight we reach a precipice where, at the edge of the glacier, we have to camp. There is a ledge barely three feet wide where we settle down. It surmounts a steep descent for the morrow. Our feet are icy cold and our boots freeze. For thirteen hours we have been tramping through deep, wet snow and they are absolutely soaked. Paldin and Kirken take shelter beneath an overhanging rock. Shovelling away some snow from the ledge, we pitch the little tent, and Dewan Singh and I crawl into it. The precipice below extends for about 1600 feet down to another glacier, and we are camped at about 17,700 feet. Dewan Singh disappears into my sleeping-sack. With meta fuel I boil a little snow-water to make tea. Both Paldin and Kirken are suffering from headache. A warm drink will do them good, but they are too tired to prepare it for themselves. This job done, I find that I really have not enough room in the tiny tent. Still, Dewan Singh has done his best, rolling himself into a ball as far as he can get. I lie with my head in the open, contemplating the moonlit sky. Then I fall asleep, to awaken after a while in thick mist. Dewan Singh crackles as he stirs in his frozen sleeping-sack. Below our ledge a great block of stone thunders down the precipice.

"August 28th. Cautiously, our limbs stiff with cold, we clamber down the precipice. None of the stones is a secure foothold, and everything is covered with recent snow. Kirken, alarmed or tired out, is scarcely able to move. Dewan Singh relieves him of the tent.

"On the lower lateral glacier there are not many crevasses, and they can all be crossed on solid snow-bridges. On a bare patch of rubble, where there is no snow lying, at 15,750 feet, we find cactus-like saussureas, growing to a height of twelve inches, and covered with felted hairs. Soon we get below the recent snowfall, and at 14,750 feet have to make our way through thickets of alpine roses. The crossing of the lower part of the Pindari Glacier is another difficulty, and takes up a good deal of time, for the going is very rough. Then we reach the lower end of the glacier, and immediately, at 12,000 feet, enter a thick mountain forest consisting of gnarled branches hung with long beard-moss, and in places we find almost impenetrable undergrowth of rhododendrons as tall as ourselves. Still, there is a good path, so we make excellent progress in spite of this magnificent wilderness. At 10,800 feet we hear the first cicada chirping, and 600 feet lower down we come to fir-trees, with dense bamboo as undergrowth. We camp when we get down to about 10,000 feet, surrounded by a wilderness of mountain jungle, consisting of birches, maples, yews, and pines, with vast quantities of ferns, rhododendrons, and bamboos growing among them. Through the steep gorge the Pindari runs as a roaring torrent.

Its milky, glacial water is making for the Ganges, for we have now found our way to the eastern head-waters of this sacred stream. To-day we have come down 8200 feet to reach our camp. The atmosphere is warm. We can no longer sleep in our thick sacks, rolling ourselves in blankets and using the sacks as mattresses. Large bats flutter round our camp. ▪

Louis Owens

FROM *Wolfsong* (1991)

Once outside the clump of trees, he looked out over the timber that thickened toward the river. Stretched below him was the upper reach of the Stehemish drainage, a wide vee of timber that covered the Great Fill where the glaciers had ground down the mountain and silted a deep plateau across the valley. Runouts cut from several glaciers on this side of the big mountain and formed sharp ridges crumbling into gray streams. Then the streams came together and carved a deep trench out of the Great Fill, and that was the Stehemish River which began in the ash-gray froth and the waste of the mountain and whitened on its way down the valley toward Forks more than forty miles away. Tall, old-growth timber tilted and fell off the crumbling edges of the fill and slid down to the streams to form logjams that flushed out with each spring's runoff and became the battered, polished logs that lined the banks and made new impasses as far down as the meeting between Sauk and Stehemish.

Only the tops of the glaciers showed above the ridges the glaciers had carved, the uppermost regions of the living ice sheets rising together toward the summit of the mountain that erupted in white light over the drainage. Dakobed, which the Stehemish called the mother mountain and the whites called Glacier Peak, towered above everything on the horizon, the mountain under whose ice Coyote had loved Goat Woman and brought the roof of the world tumbling in. On the slopes below him the timber grew even thicker as it climbed the sides of Miner's Creek, the creek named for the men who'd been prying and tapping like burglars at the ridge for half a century. ■

John Muir

FROM *The Yosemite* (1912)

The most striking and attractive of the glacial phenomena in the upper Yosemite region are the polished glacier pavements, because they are so beautiful, and their beauty is of so rare a kind, so unlike any portion of the loose, deeply weathered lowlands where people make homes and earn their bread. They are simply flat or gently undulating areas of hard resisting granite, which present the unchanged surface upon which with enormous pressure the ancient glaciers flowed. They are found in mostly perfect condition in the subalpine region, at an elevation of from eight thousand to nine thousand feet. Some are miles in extent, only slightly interrupted by spots that have given way to the weather, while the best preserved portions reflect the sunbeams like calm water or glass, and shine as if polished afresh every day, notwithstanding they have been exposed to corroding rains, dew, frost, and snow measureless thousands of years.

The attention of wandering hunters and prospectors, who see so many mountain wonders, is seldom commanded by other glacial phenomena, moraines however regular and artificial-looking, canyons however deep or strangely modeled, rocks however high; but when they come to these shining pavements they stop and stare in wondering admiration, kneel again and again to examine the brightest spot, and try hard to account for their mysterious shining smoothness. They may have seen the winter avalanches of snow descending in awful majesty through the woods, scouring the rocks and sweeping away like weeds the trees that stood in their way, but conclude that this cannot be the work of avalanches, because the scratches and fine polished striae show that the agent, whatever it was, moved along the sides of high rocks and ridges and up over the tops of them as well as down their slopes. Neither can they see how water may possibly have been the agent, for they find the same strange polish upon ridges and domes thousands of feet above the reach of any conceivable flood. Of all the agents of whose work they know anything, only the wind seems capable of moving across the face of the country in the directions indicated by the scratches and grooves. The Indian name of Lake Tenaya is "Pyweak"—the lake of shining rocks. One of the Yosemite tribe, Indian Tom, came to me and asked if I could tell him what had made the Tenaya rocks so smooth. Even dogs and horses, when first led up the mountains, study geology to this extent that they gaze wonderingly at the strange brightness of the ground and smell it, and place their feet cautiously upon it as if afraid of falling or sinking.

In the production of this admirable hard finish, the glaciers in many places flowed with a pressure of more than a thousand tons to the square yard, planing down granite, slate, and quartz alike, and bringing out the veins and crystals of the rocks with beautiful distinctness. Over large areas below the surface of the Tuolumne and Merced the granite is porphyritic; feldspar crystals an inch or two in length in many places form the greater part of the rock, and these, when planed off level with the general surface, give rise to a beautiful mosaic on which the happy sunbeams plash and glow in passionate enthusiasm. Here lie the brightest of all the Sierra landscapes. The range both to the north and south of this region was, perhaps, glaciated about as heavily, but because the rocks are less resisting, their polished surfaces have mostly given way to the weather, leaving only small imperfect patches. The lowest remnants of the old glacial surface occur at an elevation of from 3000 to 5000 feet above sea-level, twenty to thirty miles below the axis of the range. The short, steeply inclined canyons of the eastern flank also contain enduring, brilliantly striated and polished rocks, but these are less magnificent than those of the broad western flank. ▪

8 Wind and Desert

While he spoke we scoured along the dazzling plain,
now nearly bare of trees, and turning slowly softer under
foot. At first it had been grey shingle, packed like gravel.
Then the sand increased and the stones grew rarer, till
we could distinguish the colors of the separate flakes,
porphyry, green schist, basalt. At last it was nearly pure
white sand, under which lay a harder stratum. Such
going was like a pile-carpet for our camels' running.
The particles of sand were clean and polished, and
caught the blaze of sun like little diamonds in a reflection so fierce, that after a while I could not endure it.

—T. E. Lawrence, *Seven Pillars of Wisdom*

SAHARA, NAMIB, KALAHARI, ATACAMA, TAKLA MAKAN, ARABIAN, GREAT Australian. In the American Southwest: Mojave, Sonoran, Chihuahuan, Great Basin. Such desert names might summon images of silence, solitude, expansiveness, and star-filled nights. These lands of low and irregular rainfall, high rates of evaporation, and extreme temperatures include some of Earth's most extensive and remote regions, covering about one-seventh of the planet's surface.

The location and distribution of deserts on Earth is not haphazard. Because of global atmospheric circulation, many deserts lie in two discontinuous subtropical high-pressure belts within fifteen to thirty-five degrees north and south latitudes, where drying and warming air descends, rainfall is scarce, and winds are light and capricious. These latitudes, particularly in the oceans, are also known as the "Horse Latitudes," a term that may have originated in Europe's transocean sailing days because the slowness of crossing these becalmed regions forced many a ship's crew, facing water and food shortages, to throw horses overboard. Other deserts in continental interiors may be caused by remoteness from oceanic moisture sources or because they lie in part or wholly in rain shadows cast by mountains and thus are blocked from moisture-bearing winds. Finally, the high latitude regions of the Arctic and Antarctic are low-precipitation, low-evaporation "cold deserts."

For example, the arid lands of the American Southwest—the Mojave, Sonoran, Chihuahuan, and Great Basin deserts—are formed from the complex interplay of desert-producing factors, including latitude and the work of descending, dry air currents; long rain shadows cast by the Sierra Nevada and other mountains and high plateaus on air masses brought in from the Pacific by prevailing westerly winds; distance from ocean moisture sources; and interacting air masses over the gulfs of California and Mexico.

In all deserts, aridity and erosion have conspired to expose land without the vegetative cover of humid regions. There, Earth's composition and structure, and the relationships between them, stand out. In deserts, strong winds are significant agents of erosion and deposition, moving enormous quantities of dust, silt, and sand across vast areas, abrading and sandblasting landscapes in their path, and depositing spectacular dune fields.

One significant human narrative is the story of "desert" encounters. We can trace the evolution of the word *desert* as a descriptor of any wild, uninhabited, and uncultivated tract or desolate region to a term referring to an arid, barren area. In Judeo-Christian tradition, *desert* did carry the original meaning of wilderness or deserted area, and in the world described by the Bible, desert was associated with arid lands.[1] Historian Patricia Limerick notes in *Desert Passages* (1989) that as the word's use evolved in Europe, its association with emptiness remained. Early English colonists commonly described their "new world" of eastern North America as deserted. John Smith wrote of Virginia "the most of this country, though desert, yet exceeding fertile," and Puritan leader Cotton Mather described New England as a "Squalid, horrid American desert."[2]

The location and identity of "desert" in North America changed through the nineteenth century as Euro-Americans, in their westward movement, encountered different lands through exploration, overland migration, and settlement. The connection with aridity was partly restored as *desert* was applied to lands beyond the Appalachian Mountains, first to the sub-humid and semi-arid grasslands of the midcontinent (the "Great American Desert"), and then to the true drylands farther west. The association of these lands with vacancy or emptiness, however, was faulty because neither the east coast nor the region farther west was uninhabited or unused. By year-long occupation, and seasonal presence and mobility, Native American peoples lived and invested themselves in, and changed, those lands long before the arrival of Europeans.

In this chapter we include writings on the sources and lives of wind, and desert writings of imagination and experience, from the African Sahara and Arabian Peninsula, from the Namib, and from the Mojave and Sonora. From the book *Wind: How the Flow of Air Has Shaped Life, Myth, and the Land* (1998), Jan DeBlieu considers Diné (Navajo) concepts of wind, and how wind structures the world and human thought. From Michael Ondaatje's *The English Patient* (1992), we include the imaginative observations by the title character of the winds and dust storms he has noted around the Mediterranean, North Africa, and beyond.

From German geologist Hans Cloos's *Conversations with the Earth* (1953), we include a selection that describes the "stony pastures" of sandstorms, with rocks blasted and abraded by wind-carried sands. These rocks often comprise desert pavements where the land surface is armored by a close-packed arrangement of stones, sand and silt having been carried away by the wind.

In the excerpt from *Sahara Unveiled* (1996), William Langewiesche considers the movement and patterns of Saharan sand and the pioneering work by British officer R. A. Bagnold on the physics of blown sand. Poet, essayist, and novelist Audre Lorde (1934–1992) reflects in other ways on sand in her poem "Sahara" from *The Black Unicorn* (1978). In response to that collection, poet Adrienne Rich commented that "Lorde writes as a Black woman, a mother, a daughter, a Lesbian, a feminist, a visionary; poems of elemental wildness and healing, nightmare and lucidity."

Young German geologists Henno Martin and Hermann Korn, with their dog Otto, sought shelter in the Namib Desert in 1939 "in order to escape the madness of the Second World War" and internment. They survived there in hiding for two-and-a-half years until illness required they turn themselves in to South African authorities. The excerpt from Martin's book *The Sheltering Desert* gives a sense of the desert's abruptly changing nature with the rare coming of rain.

By the turn of the twentieth century, deserts of the American Southwest found a voice in Euro-American writers such as Mary Austin (1868–1934), one of the first truly western natural-history writers in the United States. Like John Muir and his writing of the Sierra Nevada, Austin adopted, described, and celebrated the desert as animate at a time when the harshness of the region still repelled many. In her first book, *The Land of Little Rain* (1903), Austin wrote from experience about the natural and cultural landscapes of southeastern California. This work describes with intimate detail the eastern Sierra Nevada, Owens Valley, Death Valley, and other parts of the Mojave Desert, and the people who lived and came to live there—the Paiute, Shoshone, and Latino and Anglo settlers. We include an excerpt from the first chapter of her book.

Ofelia Zepeda, a member of the Tohono O'odham nation, writes of family memory, story, and Sonoran desert rhythms in "Wind" from *Ocean Power* (1995). A linguist, poet, editor, and former MacArthur Foundation Fellow, she works to maintain and preserve indigenous languages and to revitalize tribal communities and cultures.

By the middle of the twentieth century, southwestern deserts also came to be sites of military weapons testing. There, physicists and other scientists developed the first nuclear weapons during World War II for the Manhattan Project. Our chapter ends with author Ray Gonzalez's reflection on visits to White Sands, New Mexico, from *Memory Fever.* Trinity, the first atomic bomb, was detonated near White Sands on July 16, 1945.

NOTES

1. See George H. Williams, *Wilderness and Paradise in Christian Thought: The Biblical Experience of the Desert in the History of Christianity and the Paradise Theme in the Theological Ideal of the University* (New York: Harper and Brothers, 1962); and Patricia N. Limerick, *Desert Passages: Encounters with the American Deserts* (Niwot: University Press of Colorado, 1989).

2. George R. Stewart, *Names on the Land: A Historical Account of Place-Naming in the United States,* 3rd ed. (Boston: Houghton Mifflin Company, 1967); and George H. Williams, *Wilderness and Paradise in Christian Thought: The Biblical Experience of the Desert in the History of Christianity and the Paradise Theme in the Theological Ideal of the University* (New York: Harper and Brothers, 1962).

Jan DeBlieu

FROM *Wind: How the Flow of Air Has Shaped Life, Myth, and the Land* (1998)

Deep in the red-rock, breeze-chiseled country of the American Southwest live a people whose religion may be more steeped in wind than any other in the world. The Diné, or Navajo, believe humans are brought to life by *nilch'i*, the Holy Wind, which leaves whorls on the tips of our fingers and toes, whispers words of advice in our ears, and dictates the number of days we spend on earth.

For his 1981 book *Holy Wind in Navajo Philosophy*, James Kale McNeley interviewed ten Diné about the various powers attributed to the winds. Eight of the men interviewed were elderly singers who performed the ceremonials and healing chants central to the Navajo religion, including Windway, the rite used to cure ills caused by winds, snakes, cactuses, clouds, the sun, and the moon. This gives their words special weight, since singers are charged with preserving tribal tradition.

Unlike the Western concept of wind, the Diné wind is a spiritual force, one of the holy beings. The Diné creation stories begin in an underworld with the "misting up" of lights of different hues from the horizons—in the east, a light as white as the edge of dawn; one blue like the sky at midday in the south; in the west the yellow glow of twilight; and a deep black beacon in the north. Then came the Wind. "The Wind has given men and creatures strength ever since," a Navajo man told Father Berad Haile in 1933, "for at the beginning they were shrunken and flabby until it inflated them, and the Wind was creation's first food."

Not only was Wind the source of life and breath, but it bestowed on humans (as in the biblical Genesis) the power of thought. At first all the tribes wandered aimlessly through the underworld, incapable of making plans, until they encountered Wind in the form of a human. "I will see for you," Wind told First Man, First Woman, Talking God, and Calling God. "I know about what is in this Earth and what is on it. I am Wind!" The

Peoples could not talk, but Wind taught them how. He positioned himself in the folds of their ears so he could always be close to them to act as conscience and guide. And the Peoples found that when Wind told them something would come to pass, his prediction invariably came true.

In time several Holy People climbed up to the surface of the Earth through twelve reeds. Wind emerged with them in the form of four cardinal breezes, two male and two female. These came to serve as the breath of the sacred mountains that rim the horizon of the Navajo cosmos. The most holy of these, a female wind, went to live in the east. At the same time, two winds were born of the earth, two of water, and two of clouds. These met and mingled, and "six winds [were] formed above and six below," McNeley writes. "We live between these, . . . all of which affect us and some of which cause difficulties and sickness. . . . There is only one Wind, [but] it has twelve names."

After the emergence of the Holy People from the underworld, Earth Surface People were made from primary ingredients such as soil, lightning, and water or, according to another version, from corn mixed with jewels like turquoise and white shell. These elements gave the Earth Surface People substance, but a holy breeze known as Little Wind or Wind's Child gives them breath and helps them stand erect. A person's posture, balance, and ability to speak all are gifts of the winds dwelling within him. "It is only by means of Wind that we talk. It exists at the tip of our tongues," a singer told McNeley. Winds also "stick out" from the soft spots on the tops of our heads and curving lines on our fingers and toes. "These whorls at the tips of our toes hold us to the Earth. Those at our fingertips hold us to the Sky. Because of these, we do not fall when we move about."

When a child in conceived, it takes a wind from its mother and another from its father. These lie on top of each other—as the winds of early creation are said to have done—and merge to become the child's own. Four months after conception this "Wind that stands within one" causes the child's first movements. At birth, with its first breath, the child receives another wind, sent from a holy source, that will guide his or her life. Early missionaries to the Navajo believed that the internal wind resembled a soul, but in fact the concept is much more complex and not easily translated into the vernacular of Western culture. The inner wind draws constantly from other winds and attaches the person to the entire swirling, holy atmosphere. "That within us stands from our mouth downward, it seems," a singer said. "We breathe by it. We live by it. It moves all parts, even our hearts." ▪

Michael Ondaatje

FROM *The English Patient* (1992)

There is a whirlwind in southern Morocco, the *aajej*, against which the fellahin defend themselves with knives. There is the *africo*, which has at times reached into the city of Rome. The *alm*, a fall wind out of Yugoslavia. The *arifi*, also christened *aref* or *rifi*, which scorches with numerous tongues. These are permanent winds that live in the present tense.

There are other, less constant winds that change direction, that can knock down horse and rider and realign themselves anticlockwise. The *bist roz* leaps into Afghanistan for 170 days—burying villages. There is the hot, dry *ghibli* from Tunis, which rolls and rolls and produces a nervous condition. The *haboob*—a Sudan dust storm that dresses in bright yellow walls a thousand meters high and is followed by rain. The *harmattan*, which blows and eventually drowns itself into the Atlantic. *Imbat*, a sea breeze in North Africa. Some winds that just sigh towards the sky. Night dust storms that come with the cold. The *khamsin*, a dust in Egypt from March to May, named after the Arabic word for "fifty," blooming for fifty days—the ninth plague of Egypt. The *datoo* out of Gibraltar, which carries fragrance.

There is also the ———, the secret wind of the desert, whose name was erased by a king after his son died within it. And the *nafhat*—a blast out of Arabia. The *mezzar-ifoullousen*—a violent and cold southwesterly known to Berbers as "that which plucks the fowls." The *beshabar*, a black and dry northeasterly out of the Caucasus, "black wind." The *Samiel* from Turkey, "poison and wind," used often in battle. As well as the other "poison winds," the *simoom*, of North Africa, and the *solano*, whose dust plucks off rare petals, causing giddiness.

Other, private winds.

Travelling along the ground like a flood. Blasting off paint, throwing down telephone poles, transporting stones and statue heads. The *harmattan* blows across the Sahara filled with red dust, dust as fire, as flour, entering and coagulating in the locks of rifles. Mariners called this red wind the "sea of darkness." Red sand fogs out of the Sahara were deposited as far north as Cornwall and Devon, producing showers of mud so great this was also mistaken for blood. "Blood rains were widely reported in Portugal and Spain in 1901."

There are always millions of tons of dust in the air, just as there are millions of cubes of air in the earth and more living flesh in the soil (worms, beetles, underground creatures) than there is grazing and existing on it. Herodotus records the death of various armies engulfed in the *simoom* who were never seen again. One nation was "so enraged by this evil wind that they declared war on it and marched out in full battle array, only to be rapidly and completely interred."

Dust storms in three shapes. The whirl. The column. The sheet. In the first the horizon is lost. In the second you are surrounded by "waltzing Ginns." The third, the sheet, is "copper-tinted. Nature seems to be on fire." ▪

Hans Cloos

FROM *Conversations with the Earth* (1953)
   translated from the German by E. B. Garside

That evening I stood on the gneiss hill—Diamond Mountain—above the
bay. The wind had quieted down, but in the silence the effects of its erosive power were all the more noticeable. My feet did not tread ordinary
stone, but a hard granular and porous sponge, full of large holes. The
ground sounded hollow. Loose pieces were light and could be picked up
without effort.

These were the stony pastures of the sandstorm. Here, for almost nine
months of the year without a break, the wind devours the land. It gnaws
away the rocks as hungry goats gnaw harsh grasses and thorny bushes.
Below, where the gray rock is ringed with a surging collar of foam, it vanishes under the surface of the sea. Here is the never-idle workshop of the
wind. The surf grinds the rocks, sieves and sorts the flour, retains the
coarsest and finest, and turns the medium sizes over to the wind for its
work on the land. Armed with this glass-hard quartz-shot, the wind cease-
lessly pelts the mild slate and the waxy soft limestone, the hard granite and
its schistose, somewhat less durable brother, the gneiss. Quickly, as if
melted, the soft rocks disappear. In the resulting furrows and channels the
stream of sand branches, thickens, drills, and bores its way forward.

The sand quickly eats up the leafy mica in the granites and gneisses,
devours the feldspars more slowly, and adds the quartz which is left over
to its supply of abrasives.

Woe to the rock whose protective crust is bored through, even at only
one small spot. The wind whirls in the depression, and presently a cave is
dug which steadily deepens and widens. Eventually the rock becomes a
sponge. The corroded skin crumbles and collapses, leaving the debris to
gravity and to further disintegration while exposing fresh layers to the
voracious attacker.

Something strange happens to round pebbles which have been abandoned to the wind by waters that have dried up long ago. They are sliced

obliquely, as a potato is sliced with a knife, often several times in different directions, so that the circle or ellipse of the original circumference becomes a polygon, the ball of a polyhedron. If, in addition, they are brightly colored, these three- or four-faceted stones lie around like ornaments and can be gathered like flowers, to be placed on a table at home. They are small but reliable records of a world in which real flowers cannot exist; they give unwilling evidence to us of the climate of former geological periods, just as they will tell future generations about the climate of today.

The sculptor of the desert creates his most decorative forms when he hits directly against steep walls of soft rock with interspersed harder enclosures. These harder parts are left while the softer ones recede. Thus cones remain which point into the wind and are tipped with a grain of quartz, a pebble, or a fossilized shell. They become rock pyramids that resemble the earth-pillars, near Bozen in Austria, which the rain has washed out of the moraine of clay, sand, and glacial boulders. But the desert cones are horizontal, parting the wind as a bird's head parts it, or as it is cleaved by the leading edge of a plane wing, the nose of a bullet, or the radiator of an automobile.

The sand blast effects on the gneiss hill above Lüderitzbucht are seen only in the spongily corroded, honeycombed stone surfaces. I took many large pieces with me. Rescued from death by sand blast, they now lie either in museum collections or on my desk. If I should turn them over to the moist airs of Germany, the rounded walls between the holes and hollows would weather away and be destroyed, and after a time the grotesque sculptures would collapse into miserable lumps of clay. ■

William Langewiesche

FROM "The Physics of Blown Sand" in *Sahara Unveiled* (1996)

I stayed on in El Oued and in the early hours walked south along a crumbling paved road under assault from the sand sea. The morning was bright and hot, and the dunes carved crisp lines against the sky. I passed a turbaned man on a donkey carrying empty gas cans into town. There was no other traffic. The minarets of El Oued disappeared behind me. The road led eventually to a village, or what was left of it. It was a village that had been mostly buried in drifting sand. The corners and roofs of stone structures still showed, but only three houses remained inhabitable, and from the evidence of digging around them, they, too, were threatened.

I drank at a well with a rope and bucket. There was no farming here. The only sign of industry was a freestanding stone oven, a baker's oven, against which palm wood had been stacked. The wind blew sand, but otherwise nothing stirred. Two men sat in the shadow of a wall by a fire on which they had placed a blackened pot. They motioned me over and offered me tea in a dirty glass. The men were older than I, bearded and thin, and had no work. We spent a few hours together. They pointed to where the school lay buried, and to where most of the village stood beneath the sands. I asked them the details of its burial.

They said the sands are fickle. Dunes may drift for decades in one direction, or not drift at all, then suddenly turn and consume you. Consumption by the sand is like other forms of terminal illness: it starts so gently that at first you don't worry. One day the grains begin to accumulate against your walls. You've seen the grains before, and naturally assume that a change in the wind will carry them away. But this time the wind does not change, and the illness persists. Over weeks or longer, the sand grows. You fight back with a shovel, and manage to keep your walls clear. Fighting back feels good and gives you something to do. But the grains never let up, and one morning while shoveling you realize that the dunes have moved closer. You enlist your sons and brothers. But eventually the land around your house swells with sand, and you begin standing on sand to shovel sand. Finally no amount of digging will clear your walls. The dunes tower above you, and send sand sheets cascading down their advancing slip faces. You have to gather your belongings and flee.

But your house is your heritage, and you would like somehow to preserve it. As the dunes bear down on it they will collapse the walls. The defense is again the Saharan acceptance of destiny: having lost the fight against the sand, you must now invite it in. Sleeping on the sand, covering your floors with it for all these years, helped prepare you mentally. But shoveling in the sand is not enough. Your last act is to break out the windows, take off the doors, and knock holes in the roof. You allow the wind to work for you. If it succeeds, and fills your house, the walls will stand. Then in a hundred years, when the wind requires it, the dunes will drift on and uncover the village. Your descendants will bless God and his Prophet. They will not care that you were thin and poor and had no work. They will remember you as a man at peace with his world. The desert takes away but also delivers.

I left the men to their contemplation, and climbed out over the dune that had engulfed their village. From its crest I discovered a valley two hundred feet below, where the desert floor was exposed and a stretch of blacktop emerged from the sand. The road was not on the map. It lay beyond the village and ran south toward the empty center of the Eastern Erg. I thought it might lead to an old settlement, perhaps one that had been uncovered by the wind, and I set off to follow it.

I should have been more careful. I was traveling too lightly, with neither a hat nor water nor any enduring sense of direction. The road kept turning, diving into sand, reemerging. Eventually it ran under a mountainous dune and disappeared entirely. I climbed that dune, and a string of the highest ones beyond it, and knew even as I proceeded that I had gone too far. The dunes were like giant starfish, covered by ripples, linking curved tentacles to form lines. In all directions, the erg stretched to the horizons in a confusion of sand.

This was the landscape that inspired the British officer Ralph A. Bagnold, history's closest observer of Saharan sands. Bagnold was an English gentleman of the old school. He fought in the trenches of Flanders during World War I, then earned an honors degree in engineering from Cambridge, and later re-enlisted in the British Army for overseas assignment. While stationed in Egypt and India between 1929 and 1934, he led expeditions in modified Fords to explore the sand seas of Libya. These were big places in need of understanding. One *erg* alone was the size of all France.

Bagnold had a strong and inquiring mind. He marveled at the desert's patterns, saw magic in the dunes, and wanted it all explained. To his surprise he found that scientific knowledge was as yet merely descriptive: dune shapes had been catalogued, but little was understood about the processes involved in their formation. Bagnold set out to understand for himself. He went back to England, retired from the army, hammered together a personal wind tunnel, and began a series of meticulous experiments with blown sand. He considered himself to be a dabbler, a tinkerer, an amateur scientist. But his research resulted in the publication, in 1941, of a small masterpiece of scientific exploration: *The Physics of Blown Sand and Desert Dunes*. It was a treatment so rigorous, and so pleasantly written, that it remains the standard today. Throughout it, Bagnold never lost his wonder. He wrote:

> Here, instead of finding chaos and disorder, the observer never fails to be amazed at a simplicity of form, an exactitude of repetition and a geometric order unknown in nature on a scale larger than that of crystalline structure. In places vast accumulations of sand weighing millions of tons move inexorably, in regular formation, over the surface of the country, growing, retaining their shape, even breeding, in a manner which, by its grotesque imitation of life, is vaguely disturbing to the imaginative mind.

Bagnold's genius was his ability to think grain by grain. He defined sand as a rock particle small enough to be moved by the wind, yet not so small that, like dust, it can float indefinitely in suspension—and he proceeded from there, exploring the movement of each grain. He did his best work on that level, in a laboratory far from the desert. But he was never a tedious man. He understood the power of multiplication. And when he returned to the Sahara, and stood as I did on the crests of the great ergs, he found in these accumulations his truest companions. Just before his death, in May 1990, he wrote a short memoir—an unintentionally sad remembrance of a strong life. He wrote about two world wars, about great men he had known, and about his beloved family. But again he wrote best about the sand. Bagnold's health was declining. It is a measure of the man that when he described the dunes' ability to heal themselves, his writing remained free of longing. ■

Audre Lorde

FROM *The Black Unicorn* (1978)

SAHARA

High
above this desert
I am
becoming
absorbed.

Plateaus of sand
dendrites of sand
continents and islands and waddys
of sand
tongue sand
wrinkle sand
mountain sand
coasts of sand
pimples and pustules and macula of sand
snot all over your face from sneezing sand
dry lakes of sand
buried pools of sand
moon craters of sand
Get your "I've had too much of people"
out of here sand.

My own place sand
never another place sand
punishments of sand
hosannahs of sand
Epiphanies of sand
crevasses of sand
mother of sand
I've been here a long time sand

string sand
spaghetti sand
cat's cradle ring-a-levio sand
army of trees sand
jungle of sand
grief of sand
subterranean treasure sand
moonglade sand
male sand
terrifying sand

Will I never get out of here sand
open and closed sand
curvatures of sand
nipples of sand
hard erected bosoms of sand
clouds quick and heavy and
desperate sand
thick veil over my face sand
sun is my lover sand
footprints of the time on sand
navel sand
elbow sand
play hopscotch through the labyrinth sand
I have spread myself sand
I have grown harsh and flat
against you sand
glass sand
fire sand
malachite and gold diamond sand
cloisonné coal sand
filagree silver sand
granite and marble and ivory sand

Hey you come here and she came sand
I will endure sand
I will resist sand
I am tired of no
all the time sand
I too will unmask my dark
hard rock sand. ▪

Henno Martin

"Desert Rain" in *The Sheltering Desert* (1957)

January passed into February, and the sun still dominated the sky unchallenged. Day after day death poured down on the earth, its scythe a sheaf of burning rays, and its harvest as generous as in war. The air trembled over the plains as it does over a hot oven, and the gramadulla gorges yawned like the portals of hell. Hyaenas and leopards had easy game with their exhausted prey, and in the shade of the few bushes and trees myriads of sand ticks waited for the wretched animals.

At last, on February 14th, great white clouds billowed up over the mountains. In the evening they retreated once more before the west wind, but the next day they sailed overhead like fully rigged ships, and by evening thunder was rolling heavily over the Namib. Humid masses of air from the Indian Ocean had swept across the African continent and reached the edge of the desert.

The next day we began to prepare ourselves a shelter from the expected rain, because the wattled roof of our hut was not water-tight. About 130 feet up we found a long narrow space between fallen rock and the mountain side. We built a stone wall on the eastern side and made a roof with our tarpaulin, weighting it down with stones. In this way we had a narrow shelter which would keep out the rain. All our perishable goods were in the driving cabin of the lorry as usual and the lorry itself was safely parked under an overhanging rock face.

From our point of vantage we marvelled at billowing clouds bigger than we had ever seen before. Here and there deep blue shadows were already creeping over the rock ridges and the gorges. At twelve o'clock a dark-grey mass moved slowly forward over the Quabis range, gathering momentum as it advanced; then it tipped over into the plains, rolling forward like lava from a volcano. Where it touched the plains great veils of sand rose up at once and were swept off into the Namib like reddish ghosts. We were still staring after those whirls of sand when amidst a ceaseless rolling of thunder the great cloud collapsed in a grey swirl and disappeared so completely that not even a wisp of mist was left. The whole thing had lasted perhaps a quarter of an hour, and an hour later we heard the roaring of water rushing down the main gorge. We ran quickly to see it as it swirled along, a tumultuous frothing bore of brown water over six feet high, flattening the tough bushes in its path, uprooting trees and tossing them into the air like matchsticks. It seemed almost incredible that such a short downpour could produce such a volume of water.

Clouds were rolling up higher into the sky now and obscuring the sun. By three o'clock it was dark and eerily quiet. Suddenly there was a rumbling of thunder and it did not stop.

"Quick!" I shouted. "We're going to get it."

And I ran back to get the wireless into safety. After that everything went very quickly. A powerful gust of wind bent the bushes flat, uprooted a small tree and flung it at our feet, and almost blew us over. The first drops fell when we were making the second journey with our bedding. On the way back the rain was pelting down so violently that we could hardly breathe; flashes of lightning lit up the grey curtain of water, and the thunder rolled and reverberated incessantly. Within minutes the water was rolling down the sides of the valley into the river, raising the level of the rushing torrent so that it now overflowed its banks and swirled into our hut. In the nick of time I rescued our small battery from knee-deep water, and as we worked and slaved to save our things before the hut fell in, as it threatened to do at any moment, the water came up to our thighs. At the gorge bend there was a roaring waterfall now and débris had piled up at every bush. The swirling water was slashing our legs with briars and I trod on a thorn and drove it into my foot. Hermann brought up the last of our belongings and we crouched at the entrance of our shelter and stared at each other in alarm: "Where was Otto?"

We called to him as loud as we could, but there was no reply. We looked in all directions, but we could see nothing but water. Finally we discovered him sitting at the back of our shelter behind the wireless, trembling and looking very frightened. In our great relief we laughed heartily at the pitiful sight of him. Gradually the clouds broke up, became ragged and swept by; the thunder died away in the west over Namib, and the flow of the gurgling water began to ebb. The cloud-burst had lasted half an hour, and we measured four inches of rain, more than the yearly average of a good many towns in Europe.

Night fell and sheet lightning could be seen stabbing the clouds on the horizon like searchlights; and then, somewhere away to the south, the incessant roll of thunder began again. Otto got as close to Hermann as possible. I could quite understand him. I had never before in my life heard such thunder or experienced such a cloud-burst.

The drought was over and every day great towering clouds rose over the uplands, bellying up into the sky to a height of thirty thousand feet and more. On one occasion we counted from ten to fifteen of them along a front of about 120 miles. They would stand there for a while like giants in shining armor on guard over the blue mountains which mark the barrier of Africa against the desert. Then forked lightning would play around their dark base and they would begin to roll forward majestically over the mountains and down into the plains.

It was as though a good fairy had broken a spell that had rested heavily on the land, and now the scorched and battered life began to raise its head again with all the vigor and confidence of youth. Within four hours, bushes that had looked dead began to show tiny shoots of green, and in the shade of the rocks ferns began to unroll delicate light-green leaves. In twenty-four hours the evening sun breaking through a bank of clouds turned the fine new blades of springbok grass into a mist of gold-green bloom. The desert was alive everywhere: seeds that had lain dormant for years came to life and pierced the crust of the earth; almost overnight the balsam bushes covered themselves with green leaves like young birches; a delicate tracery of green creepers began to wind over the red sand; and the first yellow flowers opened their petals to the sun.

Once again we laid out a small garden, fetching the water from the overbrimming holes. Rock doves were billing and cooing from morning to night; a pair of garden-warblers nested under a piece of bark near our lorry; before long we found the speckled eggs of the quail amidst the grass and stones; and the lukewarm water of the pools swarmed with little crab-like insects whose eggs had survived the years of dryness and scorching sunshine. ■

Mary Austin

FROM *The Land of Little Rain* (1903)

East away from the Sierras, south from Panamint and Amargosa, east and south many an uncounted mile, is the Country of Lost Borders.

Ute, Paiute, Mojave, and Shoshone inhabit its frontiers, and as far into the heart of it as a man dare go. Not the law, but the land sets the limit. Desert is the name it wears upon the maps, but the Indian's is the better word. Desert is a loose term to indicate land that supports no man; whether the land can be bitted and broken to that purpose is not proven. Void of life it never is, however dry the air and villainous the soil.

This is the nature of that country. There are hills, rounded, blunt, burned, squeezed up out of chaos, chrome and vermilion painted, aspiring to the snowline. Between the hills lie high level-looking plains full of intolerable sun glare, or narrow valleys drowned in a blue haze. The hill surface is streaked with ash drift and black, unweathered lava flows. After rains water accumulates in the hollows of small closed valleys, and, evaporating, leaves hard dry levels of pure desertness that get the local name of dry lakes. Where the mountains are steep and the rains heavy, the pool is never quite dry, but dark and bitter, rimmed about with the efflorescence of alkaline deposits. A thin crust of it lies along the marsh over the vegetating area, which has neither beauty nor freshness. In the broad wastes open to the wind the sand drifts in hummocks about the stubby shrubs, and between them the soil shows saline traces. The sculpture of the hills here is more wind than water work, though the quick storms do sometimes scar them past many a year's redeeming. In all the Western desert edges there are essays in miniature at the famed, terrible Grand Cañon, to which, if you keep on long enough in this country, you will come at last.

Since this is a hill country one expects to find springs, but not to depend upon them; for when found they are often brackish and unwholesome, or maddening, slow dribbles in a thirsty soil. Here you find the hot sink of Death Valley, or high rolling districts where the air has always a tang of frost. Here are the long heavy winds and breathless calms on the tilted mesas where dust devils dance, whirling up into a wide, pale sky. Here you have no rain when all the earth cries for it, or quick downpours called cloud-bursts for violence. A land of lost rivers, with little in it to love; yet a land that once visited must be come back to inevitably. If it were not so there would be little told of it.

This is the country of three seasons. From June on to November it lies hot, still, and unbearable, sick with violent unrelieving storms; then on until April, chill, quiescent, drinking its scant rain and scanter snows; from April to the hot season again, blossoming, radiant, and seductive. These months are only approximate; later or earlier the rain-laden wind may drift up the water gate of the Colorado from the Gulf, and the land sets its seasons by the rain.

The desert floras shame us with their cheerful adaptations to the seasonal limitations. Their whole duty is to flower and fruit, and they do it hardly, or with tropical luxuriance, as the rain admits. It is recorded in the report of the Death Valley expedition that after a year of abundant rains, on the Colorado desert was found a specimen of Amaranthus ten feet high. A year later the same species in the same place matured in the drought at four inches. One hopes the land may breed like qualities in her human offspring, not tritely to "try," but to do. Seldom does the desert herb attain the full stature of the type. Extreme aridity and extreme altitude have the same dwarfing effect, so that we find in the high Sierras and in Death Valley related species in miniature that reach a comely growth in mean temperatures. Very fertile are the desert plants in expedients to prevent evaporation, turning their foliage edgewise toward the sun, growing silky hairs, exuding viscid gum. The wind, which has a long sweep, harries and helps them. It rolls up dunes about the stocky stems, encompassing and protective, and above the dunes, which may be, as with the mesquite, three times as high as a man, the blossoming twigs flourish and bear fruit.

There are many areas in the desert where drinkable water lies within a few feet of the surface, indicated by the mesquite and the bunch grass (*Sporobolus airoides*). It is this nearness of unimagined help that makes the tragedy of desert deaths. It is related that the final breakdown of that hapless party that gave Death Valley its forbidding name occurred in a locality where shallow wells would have saved them. But how were they to know that? Properly equipped it is possible to go safely across that ghastly sink, yet every year it takes its toll of death, and yet men find there sundried mummies, of whom no trace or recollection is preserved. To underestimate one's thirst, to pass a given landmark to the right or left, to find a dry spring where one looked for running water—there is no help for any of these things.

Along springs and sunken watercourses one is surprised to find such water-loving plants as grow widely in moist ground, but the true desert breeds its own kind, each in its particular habitat. The angle of the slope, the frontage of a hill, the structure of the soil determines the plant. South-looking hills are nearly bare, and the lower tree-line higher here by a thousand feet. Cañons running east and west will have one wall naked and one clothed. Around dry lakes and marshes the herbage preserves a set and orderly arrangement. Most species have well-defined areas of growth, the best index the voiceless land can give the traveler of his whereabouts.
[. . .]

If one is inclined to wonder at first how so many dwellers came to be in the loneliest land that ever came out of God's hands, what they do there and why stay, one does not wonder so much after having lived there. None other than this long brown land lays such a hold on the affections. The rainbow hills, the tender bluish mists, the luminous radiance of the spring, have the lotus charm. They trick the sense of time, so that once inhabiting there you always mean to go away without quite realizing that you have not done it. Men who have lived there, miners and cattle-men, will tell you this, not so fluently, but emphatically, cursing the land and going back to it. For one thing there is the divinest, cleanest air to be breathed anywhere in God's world. Some day the world will understand that, and the little oases on the windy tops of hills will harbor for healing its ailing, house-weary broods. There is promise there of great wealth in ores and earths, which is no wealth by reason of being so far removed from water and workable conditions, but men are bewitched by it and tempted to try the impossible.

You should hear Salty Williams tell how he used to drive eighteen and

twenty-mule teams from the borax marsh to Mojave, ninety miles, with
the trail wagon full of water barrels. Hot days the mules would go so mad
for drink that the clank of the water bucket set them into an uproar of
hideous, maimed noises, and a tangle of harness chains, while Salty would
sit on the high seat with the sun glare heavy in his eyes, dealing out curses
of pacification in a level, uninterested voice until the clamor fell off from
sheer exhaustion. There was a line of shallow graves along that road; they
used to count on dropping a man or two of every new gang of coolies
brought out in the hot season. But when he lost his swamper, smitten
without warning at the noon halt, Salty quit his job; he said it was "too
durn hot." The swamper he buried by the way with stones upon him to
keep the coyotes from digging him up, and seven years later I read the pen-
ciled lines on the pine headboard, still bright and unweathered.

But before that, driving up on the Mojave stage, I met Salty again crossing Indian Wells, his face from the high seat, tanned and ruddy as a harvest
moon, looming through the golden dust above his eighteen mules. The
land had called him.

The palpable sense of mystery in the desert air breed fables, chiefly of
lost treasure. Somewhere within its stark borders, if one believes report, is
a hill strewn with nuggets; one seamed with virgin silver; an old clayey
water-bed where Indians scooped up earth to make cooking pots and
shaped them reeking with grains of pure gold. Old miners drifting about
the desert edges, weathered into the semblance of the tawny hills, will tell
you tales like these convincingly. After a little sojourn in that land you will
believe them on your own account. It is a question whether it is not better to be bitten by the little horned snake of the desert that goes sideways
and strikes without coiling, than by the tradition of a lost mine.

And yet—and yet—is it not perhaps to satisfy expectation that one falls
into the tragic key in writing of desertness? The more you wish of it, the
more you get, and in the mean time lose much of pleasantness. In that
country which begins at the foot of the east slope of the Sierras and spreads
out by less and less lofty hill ranges toward the Great Basin, it is possible
to live with great zest, to have red blood and delicate joys, to pass and
repass about one's daily performance an area that would make an Atlantic

seaboard State, and that with no peril, and, according to our way of thought, no particular difficulty. At any rate, it was not people who went into the desert merely to write it up who invented the fabled Hassaympa, of whose waters, if any drink, they can no more see fact as naked fact, but all radiant with the color of romance. I, who must have drunk of it in my twice seven years' wanderings, am assured that it is worth while.

For all the toll the desert takes of a man it gives compensations, deep breaths, deep sleep, and the communion of the stars. It comes upon one with new force in the pauses of the night that the Chaldeans were a desert-bred people. It is hard to escape the sense of mastery as the stars move in the wide clear heavens to risings and settings unobscured. They look large and near and palpitant; as if they moved on some stately service not needful to declare. Wheeling to their stations in the sky, they make the poor world-fret of no account. Of no account you who lie out there watching, nor the lean coyote that stands off in the scrub from you and howls and howls. ▪

Ofelia Zepeda

FROM *Ocean Power* (1995)

WIND

The wind was whipping my clothes harshly around me,
slapping me,
hurting me with the roughness.
The wind was strong that evening.
It succeeded in blowing my clothes all around me.
Unlike others I revel in it.
I open my mouth and breathe it in.
It is new air,
air, coming from faraway places.
From skies untouched,
from clouds not yet formed.
I breathe in big gasps of this wind.
I think I know a secret, this is only the opening act
of what is yet to come.

I see it coming from a long distance away.
A brown wall of dust and dirt,
moving debris that is only moments old,
debris that is hundreds of years old.
All picked up in a chaotic dance.
The dust settles in my nostrils.
It clings to the moisture in my mouth.
It settles on my skin and fine hairs.

Memories of father and how he sat in front of the house
watching the wind come.
First he would smell it, then he would see it.
He would say, "Here he comes,"
much in the same way as if he saw a person on the horizon.

He would sit.
Letting the wind do with him what it will,
hitting him with pieces of sand.
Creating a fine layer all over him.
Finally when he could not stand it any longer
he would run into the house, his eyes shut,
shut against the tears getting ready to cleanse his eyes.
We all laughed at his strange appearance.
He also reveled in this wind.
This was as close as he could get to it,
to join it, to know it, to know what the wind brings.
My father would say, "Just watch, when the wind stops,
the rain will fall."

The story goes.
Wind got in trouble with the villagers.
His punishment was that he should leave the village forever.
When he received his sentence of exile
Wind went home and packed his things.
He packed his blue winds.
He packed his red winds.
He packed his black winds.
He packed his white winds.
He packed the dry winds.
He packed the wet winds.
And in doing this he took by the hand
his friend who happened to be blind,
Rain.
Together they left.
Very shortly after, the villagers found their crops began to die.
The animals disappeared,
and they were suffering from hunger and thirst.
To their horror the people realized they were wrong
in sending Wind away.

And like all epic mistakes it took epic events
to try to bring Wind back.

In the end it was a tiny tuft of down
that gave the signal that Wind was coming back.
With him was his friend, Rain.
He brought back the dry wind,
the cold wind,
the wet wind,
the cool wind,
but in his haste,
he forgot
the blue wind,
the white wind,
the red wind,
and the black wind. ▪

Ray Gonzalez

FROM "White Sands" in *Memory Fever* (1993)

It was like playing on the moon as we rolled down the white sands when we were kids. By the time we came to a stop at the bottom, we were covered in white dirt, looking like ghost children. The memory of family picnics at White Sands National Monument near Alamogordo, New Mexico, in the fifties is blurred now, but I still cling to a few vivid images of miles and miles of a white world where I had fun playing in the dunes, building white sand castles without knowing we were only a few miles from the government test site where Trinity, the first atomic bomb, was detonated July 16, 1945.

Until I studied World War II in high school, I did not know the white desert playground of the park distracted visitors from the fact that they were near the historic site. After seeing what I was reading, my mother casually told me that she and my father, along with thousands of El Pasoans, had seen a bright flash in the sky on July 16. My parents were high school students at El Paso Technical when the test bomb went off one hundred miles to the north. No one knew what the flash was until years later, but she told me that day felt very unusual. The sudden light made many people nervous because the war was still going on.

My parents had witnessed a turning point in history by seeing that bright light in the sky. As a child, I spent many weekends playing in the white sands of the future, just another kid amazed at the endless horizon of white hills and dunes, the light falling from the desert sun to wither everything in 95-degree heat, making us play harder as the light intensified.

I can't forget the first day we walked through the park museum. Besides the usual geological charts and raised maps, I saw displays of mounted white mice, white rabbits, and white coyotes—even white tarantulas. These creatures had adjusted to this white world so that they could survive in the heat and desolation of the bleached landscape. Each species evolved to take advantage of an endless camouflage. When I learned about Trinity, I wondered how many of the animals had been affected by the radiation of history. Had they truly changed color to blend into the white sand for protection and survival against predators, or was this ivory land, and its creatures, a mutation from the first blast?

Concerns over nuclear fallout were not hot topics for high school students in an isolated, west Texas town during the sixties. No one thought about it. It was not an issue. I knew nature had created white sands thousands of years ago and the evolution of white animals had nothing to do with the atomic bomb, but as I remembered that day in the museum, I sensed the whiteness of every living thing around me was connected to the darkness of the brilliant flash of 1945.

I first read Leslie Groves' eyewitness account of the Alamagordo explosion when my American History class studied the development of the atomic bombs the U.S. dropped on Hiroshima and Nagasaki. As director of the Manhattan Project, a top-secret government program, Groves described what he saw across the white sands as an intense and blinding flash of light, a tremendous ball of fire that turned into the first mushroom cloud that any of them saw. He watched as the steel tower vaporized in the 15,000 to 20,000 tons of explosives.

[. . .]

I didn't make the connection between the bomb, my parents seeing the flash, and my naive childhood in the sand until 1977, as I drove alone across southern New Mexico. I was on my way back to El Paso after living for two years in San Diego, California. I drove south past Albuquerque. But instead of taking the straight, short route to El Paso, I headed east from Las Cruces over the Organ Mountains.

The high pass took me to the eastern side of the range, directly above the vast flatness of the desert floor and White Sands, forty miles away. I don't know why I chose to return to El Paso from that direction, but driving past White Sands brought it all back. Perhaps I thought of those days because friends of mine in San Diego had mentioned their participation in antinuclear groups protesting at the nuclear power plant near San Clemente. The last demonstration had taken place a few days before I said good-bye to them and headed home.

It was an early evening in April when I headed down the long stretch of highway bordering the eastern end of White Sands. The setting sun ignited the peaks of the Organ Mountains and washed the miles of white sand with a peach-colored hue. I drove past fenced-off land whose barbed-wired barriers held small signs every few hundred feet. "Property of United States Government. No Trespassing."

A simple warning like that was enough to keep people away and draw them to the park instead. Thousands of visitors came each year to gawk at the white sand dunes, miles and miles of them. [. . .]

I have never been an antinuclear activist or protester, and have never even read books about Hiroshima, but as I drove through the area where Trinity was detonated, I thought about those childhood picnics, the white animals in the museum, and the brilliance of the sands. I discovered a certain light, a clarity of understanding, a radiance burning through the dunes of family history.

Its energy came from many sources. The innocence of childhood produced the light on those white hills. Our family outings generated the heat because they were part of growing up with a sense that the concept of family could never be broken. Those picnics on the dunes represented the idealism of a small boy growing up in the late fifties, a boy who thought the world was made up of nothing but good times.

The energy of the blinding desert also came from the fact that the U.S. government chose this area in which to conduct experiments that would change world history. If someone had interrupted my fun on the dunes to tell me that the Manhattan Project had reached its climax a few miles away, I couldn't have comprehended it. As a child, I would have thought the idea of a big bomb in the desert was kind of neat. Nothing could have destroyed my playground.

Twelve years after my solitary drive past the dunes toward home, the first test of a Star Wars laser weapon was successfully completed at White Sands. In 1989, the U.S. government finally shot down a missile with a laser. The flight and destruction of the missile took place over White Sands. Forty-four years after the first atomic flash startled the people of El Paso, another turning point had taken place over the white dunes. ▪

9 Living on Earth

. . . landscape is the most immediate medium through which we attempt to convert culturally shared dreams into palpable realities. Our actions in the world, in short, are shaped by the paradigms in our head.

—Annette Kolodny, *The Land before Her*

MYRIAD, DIVERSE LIFEWAYS AND PERSPECTIVES OF THIS EARTH CANNOT be distilled or generalized. For many peoples Earth *is* a visionary and inclusive world in which land lives as literal and spiritual home. For some, proof and evidence by scientific analysis base conceptions of the world. How large might the geologic experience be? To sense, observe, imagine Earth—in material, process, and time—is one piece, but what of how we have lived in, shaped, and marked this world? We are all hunters, gatherers, surveyors, and shapers in this open-ended journey of evolution and existence. Human history, natural history, geologic history merge into and emerge from real-world storied landscapes.

In *Hopes and Impediments* (1989), Nigerian novelist and poet Chinua Achebe wrote that "Africa is not only a geographical expression; it is also a metaphysical landscape—it is in fact a view of the world and of the whole cosmos perceived from a particular position." Human action and thought have always existed in cultural, social, geological, and ecological contexts and *positions.* The following mix of stories from Africa, Asia, Europe, and the Americas, although necessarily incomplete, offers complex and conflicting *positions* as food for thought. What is our place on Earth?

The chapter opens with the introduction to *The Way to Rainy Mountain* (1969) by N. Scott Momaday, mixed blood Kiowa author, artist, and winner of the Pulitzer Prize for fiction. This book grew out of his need for personal and cultural self-definition. As Momaday tried to retrieve remnants of Kiowa oral tradition, he realized how much the transmission of oral narratives had already deteriorated. Unable to speak the language, deprived of grandparents as living sources, he collected stories from tribal elders—to keep them alive in the imagination of modern heirs (and in printed form).

In an excerpt from *Out of the Earth,* soil scientist Daniel Hillel discusses two very different accounts of creation in the first two chapters of the Book of Genesis in the Hebrew Bible—in the second chapter of Genesis, "God Yahweh formed man out of the soil of the earth. . . ." Hillel goes on to consider other cultural associations and the importance of language, noting for example that the name "Adam," derived from the Hebrew word *adama* for soil or earth, clearly shows the tie between humans and the land.

Nikos Kazantzakis (1883–1957), author of *The Odyssey: A Modern Sequel,* *Zorba the Greek,* and *The Last Temptation of Christ,* was a novelist, columnist, poet, and travel writer. In *Report to Greco* (1965) he outlines how deeply rooted his life and work are in the natural world, in the soil of Crete.

Kofi Awoonor, Ghanaian writer and poet, as well as teacher and diplomat, noted that "landscape intertwines with the most fundamental framework of our existence." In the selection from his *The Breast of the Earth* (1975), an examination of Africa and the continent's literature, Awoonor reflects on an African conception of the sacredness and power of land.

The opening chapter of John Steinbeck's Pulitzer Prize–winning *The Grapes of Wrath* (1939) introduces the reader to the beginning of the Dust Bowl, the years of drought and human dislocation in the western Great Plains of the United States in the 1930s. Steinbeck's book goes on to examine the realities of the agrarian dream to the working man in early-mid twentieth-century America.

Leslie Marmon Silko is a mixed-blood Laguna Pueblo poet, novelist, film-maker, and former MacArthur Foundation Fellow. In an excerpt from the widely anthologized essay "Interior and Exterior Landscapes: The Pueblo Migration Stories," she describes some of the complex interrelationships in the Pueblo landscape, explaining how "the land, the sky, and all that is within them . . . includes human beings."

Then we return to David Leveson, the questioning geologist who, in *A Sense of the Earth* (1972), wrote of our need to understand the larger "geologic experience."

The book ends with part of the essay "An American Land Ethic," in which N. Scott Momaday explores a sane land ethic, believing that "once in his life a man ought to concentrate his mind upon the remembered earth."

N. Scott Momaday

FROM *The Way to Rainy Mountain* (1969)

Introduction

A single knoll rises out of the plain in Oklahoma, north and west of the Wichita Range. For my people, the Kiowas, it is an old landmark, and they gave it the name Rainy Mountain. The hardest weather in the world is there. Winter brings blizzards, hot tornadic winds arise in the spring, and in summer the prairie is an anvil's edge. The grass turns brittle and brown, and it cracks beneath your feet. There are green belts along the rivers and creeks, linear groves of hickory and pecan, willow and witch hazel. At a distance in July or August the steaming foliage seems almost to writhe in fire. Great green and yellow grasshoppers are everywhere in the tall grass, popping up like corn to sting the flesh, and tortoises crawl about on the red earth, going nowhere in the plenty of time. Loneliness is an aspect of the land. All things in the plain are isolate; there is no confusion of objects in the eye, but *one* hill or *one* tree or *one* man. To look upon that landscape in the early morning, with the sun at your back, is to lose the sense of proportion. Your imagination comes to life, and this, you think, is where Creation was begun.

I returned to Rainy Mountain in July. My grandmother had died in the spring, and I wanted to be at her grave. She had lived to be very old and at last infirm. Her only living daughter was with her when she died, and I was told that in death her face was that of a child.

I like to think of her as a child. When she was born, the Kiowas were living the last great moment of their history. For more than a hundred years they had controlled the open range from the Smoky Hill River to the Red, from the headwaters of the Canadian to the fork of the Arkansas and Cimarron. In alliance with the Comanches, they had ruled the whole of the southern Plains. War was their sacred business, and they were among the finest horsemen the world has ever known. But warfare for the Kiowas was preeminently a matter of disposition rather than of survival, and they never understood the grim, unrelenting advance of the U.S. Cavalry. When at last, divided and ill-provisioned, they were driven onto the Staked Plains in the cold rains of autumn, they fell into panic. In Palo Duro Canyon they abandoned their crucial stores to pillage and had nothing then but their lives. In order to save themselves, they surrendered to the soldiers at Fort Sill and were imprisoned in the old stone corral that now stands as a military museum. My grandmother was spared the humiliation of those high gray walls by eight or ten years, but she must have known from birth the affliction of defeat, the dark brooding of old warriors.

Her name was Aho, and she belonged to the last culture to evolve in North America. Her forebears came down from the high country in western Montana nearly three centuries ago. They were a mountain people, a mysterious tribe of hunters whose language has never been positively classified in any major group. In the late seventeenth century they began a long migration to the south and east. It was a journey toward the dawn, and it led to a golden age. Along the way the Kiowas were befriended by the Crows, who gave them the culture and religion of the Plains. They acquired horses, and their ancient nomadic spirit was suddenly free of the ground. They acquired Tai-me, the sacred Sun Dance doll, from that moment the object and symbol of their worship, and so shared in the divinity of the sun. Not least, they acquired the sense of destiny, therefore courage and pride. When they entered upon the southern Plains they had been transformed. No longer were they slaves to the simple necessity of survival; they were a lordly and dangerous society of fighters and thieves, hunters and priests of the sun. According to their origin myth, they entered the world through a hollow log. From one point of view, their migration was the fruit of an old prophecy, for indeed they emerged from a sunless world.

Although my grandmother lived out her long life in the shadow of Rainy Mountain, the immense landscape of the continental interior lay like memory in her blood. She could tell of the Crows, whom she had never seen, and of the Black Hills, where she had never been. I wanted to see in reality what she had seen more perfectly in the mind's eye, and traveled fifteen hundred miles to begin my pilgrimage.

Yellowstone, it seemed to me, was the top of the world, a region of deep lakes and dark timber, canyons and waterfalls. But, beautiful as it is, one might have the sense of confinement there. The skyline in all directions is close at hand, the high wall of the woods and deep cleavages of shade. There is a perfect freedom in the mountains, but it belongs to the eagle and the elk, the badger and the bear. The Kiowas reckoned their stature by the distance they could see, and they were bent and blind in the wilderness.

Descending eastward, the highland meadows are a stairway to the plain. In July the inland slope of the Rockies is luxuriant with flax and buckwheat, stonecrop and larkspur. The earth unfolds and the limit of the land recedes. Clusters of trees, and animals grazing far in the distance, cause the vision to reach away and wonder to build upon the mind. The sun follows a longer course in the day, and the sky is immense beyond all comparison. The great billowing clouds that sail upon it are shadows that move upon the grain like water, dividing light. Farther down, in the land of the Crows and Blackfeet, the plain is yellow. Sweet clover takes hold of the hills and bends upon itself to cover and seal the soil. There the Kiowas paused on their way; they had come to the place where they must change their lives. The sun is at home on the plains. Precisely there does it have the certain character of a god. When the Kiowas came to the land of the Crows, they could see the dark lees of the hills at dawn across the Bighorn River, the profusion of light on the grain shelves, the oldest deity ranging after the solstices. Not yet would they veer southward to the caldron of the land that lay below; they must wean their blood from the northern winter and hold the mountains a while longer in their view. They bore Tai-me in procession to the east.

A dark mist lay over the Black Hills, and the land was like iron. At the top of a ridge I caught sight of Devil's Tower upthrust against the gray sky as if in the birth of time the core of the earth had broken through its crust and the motion of the world was begun. There are things in nature that engender an awful quiet in the heart of man; Devil's Tower is one of them. Two centuries ago, because they could not do otherwise, the Kiowas made a legend at the base of the rock. My grandmother said:

> Eight children were there at play, seven sisters and their brother. Suddenly the boy was struck dumb; he trembled and began to run upon his hands and feet. His fingers became claws, and his body was

covered with fur. Directly there was a bear where the boy had been. The sisters were terrified; they ran, and the bear after them. They came to the stump of a great tree, and the tree spoke to them. It bade them climb upon it, and as they did so it began to rise into the air. The bear came to kill them, but they were just beyond its reach. It reared against the tree and scored the bark all around with its claws. The seven sisters were borne into the sky, and they became the stars of the Big Dipper.

From that moment, and so long as the legend lives, the Kiowas have kinsmen in the night sky. Whatever they were in the mountains, they could be no more. However tenuous their well-being, however much they had suffered and would suffer again, they had found a way out of the wilderness.

My grandmother had a reverence for the sun, a holy regard that now is all but gone out of mankind. There was a wariness in her, and an ancient awe. She was a Christian in her later years, but she had come a long way about, and she never forgot her birthright. As a child she had been to the Sun Dances; she had taken part in those annual rites, and by them she had learned the restoration of her people in the presence of Tai-me. ▪

Daniel Hillel

FROM *Out of the Earth* (1991)

The Hebrew Bible provides a profoundly symbolic account of the act of creation, the beginning of life on earth and the origin and role of human-kind. It describes how, after summoning up radiant energy ("Let there be light!"), the Creator imposed form and order upon the primeval chaos by separating land from water, and earth from sky. The sea and the land were then made to generate a myriad living species, and man—the presumed pinnacle of creation—was granted a privileged status in the hierarchy of life. This much of the account is known by all.

Less widely noticed is the curious fact that the first two chapters in the Book of Genesis actually give not one but two accounts of creation. Of many contradictions between the two, for us the most significant is the role assigned to humans in the scheme of life on earth. In the first chapter we read that God (called by the plural name "Elohim") decided to "make man in our own image, after our likeness, and let them rule over the fish of the sea and over the fowl of the air, and over the cattle, and over all the earth, and over every creeping thing that creepeth upon the earth." And God blessed man and woman and said unto them: "Be fruitful, and multi-ply, and *fill the earth, and conquer it*; and rule over the fish of the sea and the fowl of the air and every animal creeping over the earth." And fur-thermore God said: "Here, I have given you every herb yielding seed and every tree with fruit . . . to you it shall be for food." All this can be con-strued as a divine ordination of humans to dominate the earth and use everything on it for their own purpose.

But the act of creation and the divine injunction to man are described quite differently in the second chapter of Genesis: "God Yahweh formed man out of the soil of the earth and blew into his nostrils the breath of life, *and man became a living soul.* And God Yahweh planted a garden in Eden in the east and placed the man therein." Then comes the crucial state-ment: "God Yahweh took the man and put him in the Garden of Eden *to serve and preserve it.*" Here, man is not given license to rule over the environment and use it for his purposes alone, but—quite the contrary—is charged with the responsibility to nurture and protect God's creation.

Thus, latent in one of the main founts of Western Civilization we have two opposite perceptions of man's destiny. One is anthropocentric: man is not part of nature but set above it. His manifest destiny is to be an omnipotent master over nature, which from the outset was created for his gratification. He is endowed with the power and the right to dominate all other creatures, toward whom he has no obligations. In the words of the 115th Psalm: "The heavens are the Lord's, but the earth He hath given to the children of man." The same notion was expressed by Protagoras: "man is the measure of all things."

The other view is more earthly and modest. Man is made of soil and is given a "living soul," but no mention is made of his being "in the image of God." Man is not set above nature. Moreover, his power is constrained by duty and responsibility. Man's appointment is not an ordination but an *assignment*. The earth is not his property; he is neither its owner nor its master. Rather, man is a custodian, entrusted with the stewardship of God's garden, and he can enjoy it only on the condition that he discharge his duty faithfully. This view of humanity's role accords with the modern ecological principle that the life of every species is rooted not in separateness from nature but integration with it.

[. . .]

Readers of the Bible in translation miss much of the imagery and poetry of the evocative verbal associations in the original Hebrew. The indissoluble link between man and soil is manifest in the very name "Adam," derived from *adama*—a Hebrew noun of feminine gender meaning earth, or soil. Adam's name encapsulates man's origin and destiny: his existence and livelihood derive from the soil, to which he is tethered throughout his life and to which he is fated to return at the end of his days. Likewise, the name of Adam's mate, "Hava" (rendered "Eve" in translation) literally means "living." In the words of the Bible: "And the man called his wife Eve because she was the mother of all living." Together, therefore, Adam and Eve signify "Soil and Life."

The ancient Hebrew association of man with soil is echoed in the Latin name for man, *homo*, derived from *humus*, the stuff of life in the soil. This powerful metaphor suggests an early realization of a profound truth that humanity has since disregarded to its own detriment. Since the words "humility" and "humble" also derive from humus, it is rather ironic that we should have assigned our species so arrogant a name as *Homo sapiens sapiens* ("wise wise man"). It occurs to me, as I ponder our past and future relation to the earth, that we might consider changing our name to a more modest *Homo sapiens curans*, with the word *curans* denoting caring or caretaking, as in "curator." ("Teach us to care" was T. S. Eliot's poetic plea.) Of course, we must work to deserve the new name, even as we have not deserved the old one.

Other ancient cultures evoke powerful associations similar to those of the Hebrew Bible. In the teachings of Buddha, not only the earth itself but indeed all its life forms (even those that may seem lowliest) are spiritually sacred. To the ancient Greeks, the earth was Gaea, the great maternal goddess who, impregnated by her son and consort Uranus (god of the sky), became mother of the Titans and progenitor of all the many gods of the Greek pantheon. Among her descendants was Demeter, the goddess of agriculture, fertility, and marriage. The story of Demeter's daughter, Persephone, and—in a different context—the Egyptian god Osiris, symbolized the annual cycle of death and rebirth. In the ancient nature cults the major deity of the earth was generally feminine. The earth was seen as the source of fertility, the site of germination and regeneration, indeed the womb of life. And when the plow was invented, its use to penetrate the soil and open it for seeding seemed to simulate the very act of procreation.

Worship of the earth long predated agriculture and continued after its advent. The earth was held sacred as the embodiment of a great spirit, the creative power of the universe, manifest in all phenomena of nature. The earth spirit was believed to give shape to the features of the landscape and to regulate the seasons, the cycles of fertility, and the lives of animals and humans. Rocks, trees, mountains, springs, and caves were recognized as receptacles for this spirit, which the Romans attributed to their earth goddess, Tellus. ■

Nikos Kazantzakis

FROM *Report to Greco* (1965)
   translated from the Greek by P. A. Bien

All my forebears on my mother's side were peasants—bent over the soil, glued to the soil, their hands, feet, and minds filled with soil. They loved the land and placed all their hopes in it: over generations they and it had become one. In time of drought they grew sickly black with thirst along with it. When the first autumn rains began to rage, their bones creaked and swelled like reeds. And when they ploughed deep furrows into its womb with the share, in their breasts and thighs they re-experienced the first night they slept with their wives.

[. . .]

When a man returns to his country after many years of wandering and struggle abroad, leans against the ancestral stones, and sweeps his glance over the familiar regions so densely populated with indigenous spirits, childhood memories, and youthful longings, he breaks into a cold sweat.

The return to the ancestral soil perturbs the hearts. It is as if we were coming back from unmentionable adventures in new, forbidden regions and suddenly, there in our sojourn abroad, we sensed a weight in our hearts. What business do we have here with the pigs, eating acorns? We gaze behind us to the land we left, and sigh. Remembering the warmth, the peace, the prosperous well-being, we return like the prodigal son to the maternal breast. In me this return always caused a secret shudder, a fore-taste as though of death. It seemed I was coming back to the long-desired ancestral clay after life's jousts and prodigalities; as though darkly subter-ranean, inescapable forces had entrusted a man with the execution of spe-cific charge, and now on his return a harsh voice rose from the great bow-els of his earth and demanded, Did you carry out your charge? Give me an account of yourself!

This earthen womb knows unerringly the worth of each of her children, and the higher the soul she has fashioned, the more difficult the com-mandment she imposes on it—to save itself or its race, or the world. A man's soul is ranked by which of these commandments it is assigned, the first, the second, or the third.

It is natural that each man should see this ascent, the ascent his soul is obliged to follow, inscribed most deeply upon the soil where he was born. There is a mystical contact and understanding between this soil which fashioned us, and our souls. Just as roots send the tree the secret order to blossom and bear fruit so that they themselves may receive their justification and reach the goal of their journey, so in the same way the ancestral soil imposes difficult commandments upon the souls it has begotten. Soil and soul seem to be of the same substance, undertaking the same assault; the soul is simply maximal victory. ■

Kofi Awoonor

FROM *The Breast of the Earth* (1975)

In political conception, land forms the most solid source of power, and its alienation is the basic cause of political conflict. Land ownership is the source of conflict in East, Central, and South Africa in the convulsive relationship of white European settlers and the African peoples. Kenya became the center of this conflict at the turn of the century, with the British colonial policy that established the so-called "white highlands" in the heart of Gikuyu country.

By 1900 and during the first decades of this one, large tracts of Gikuyu land in Kenya were set aside for white settlement. It was an extension of policies already successfully tested in southern Africa, particularly in South Africa and the Rhodesias. The white settlers in Kenya took the most fertile land away from a group whose population was increasing and depended on land and agriculture for its survival. Large armies of landless men were forced to sell their labor to white farmers on their own ancestral land which had been appropriated by law. The white settler element also slowly entrenched itself politically through devious deals concluded with the metropolitan government. Thus about sixty thousand Europeans became the holders of political power in Kenya, while about seven million Africans including the Gikuyus had every single aspect of their political aspirations thwarted, frustrated, and ignored. The Africans existed only to serve the European's economic program of exploitation. Political frustration, genuine land hunger, the immediate discomfiture resulting from the Second World War, in which many Kenyans served, in Burma and the Far East, all led to the beginnings of revolt. Organizations such as the militant Gikuyu Central Association, organized by Harry Thuku in 1923, gave way to the Kenya African Union in 1945, in which old Gikuyu Central Association stalwarts like Jomo Kenyatta became the most articulate spokesmen. This later became the Kenya African National Union, the main ruling political party of Kenya today.

One of the most significant factors in the so-called Mau Mau revolt was the African's conception of land. Land was not merely an economic commodity, it was, more importantly, a sacred entity. The earth represents in most African myths the mother principle which is central to the survival and continuity of the group. That is why land is still communally held in most parts of Africa and not regarded as a commodity that can be parceled out and sold at will. In the pantheon of African gods, the Earth Goddess is an important benevolent spirit upon whom man depends for food and sustenance and to whom he returns at death. Her sanctity as the natural principle is recognized in a number of taboos and abominations which cannot be committed against her. The sanctity of the earth is recognized in ceremonies that mark the clearing of virgin forests for farming and the preparation of new plots for building. Land, thus, is the center of the community's life. The numerous first-fruits ceremonies constitute homage to the Earth Goddess, who blesses crops and yields the abundance of her bosom for her children. All human beings, the African believes, are all children of the earth; but the earth of our native soil, of our village or town, becomes the bond of the community's cohesiveness. It is for this we fight when strangers want to dislodge us. It is the place where our umbilical cord is buried, and our link stretches through her to our ancestors who were buried in her womb. It is this great sacred force that was behind the Gikuyu uprising. It informed every fierce act of heroism and bravado. ■

John Steinbeck

FROM *The Grapes of Wrath* (1939)

To the red country and part of the gray country of Oklahoma, the last rains came gently, and they did not cut the scarred earth. The plows crossed and recrossed the rivulet marks. The last rains lifted the corn quickly and scattered weed colonies and grass along the sides of the roads so that the gray country and the dark red country began to disappear under a green cover. In the last part of May the sky grew pale and the clouds that had hung in high puffs for so long in the spring were dissipated. The sun flared down on the growing corn day after day until a line of brown spread along the edge of each green bayonet. The clouds appeared, and went away, and in a while they did not try any more. The weeds grew darker green to protect themselves, and they did not spread any more. The surface of the earth crusted, a thin hard crust, and as the sky became pale, so the earth became pale, pink in the red country and white in the gray country.

In the water-cut gullies the earth dusted down in dry little streams. Gophers and ant lions started small avalanches. And as the sharp sun struck day after day, the leaves of the young corn became less stiff and erect; they bent in a curve at first, and then, as the central ribs of strength grew weak, each leaf tilted downward. Then it was June, and the sun shone more fiercely. The brown lines on the corn leaves widened and moved in on the central ribs. The weeds frayed and edged back toward their roots. The air was thin and the sky more pale; and every day the earth paled.

In the roads where the teams moved, where the wheels milled the ground and the hooves of the horses beat the ground, the dirt crust broke and the dust formed. Every moving thing lifted the dust into the air: a walking man lifted a thin layer as high as his waist, and a wagon lifted the dust as high as the fence tops, and an automobile boiled a cloud behind it. The dust was long in settling back again.

When June was half gone, the big clouds moved up out of Texas and the

Gulf, high heavy clouds, rain-heads. The men in the fields looked up at the clouds and sniffed at them and held wet fingers up to sense the wind. And the horses were nervous while the clouds were up. The rain-heads dropped a little spattering and hurried on to some other country. Behind them the sky was pale again and the sun flared. In the dust there were drop craters where the rain had fallen, and there were clean splashes on the corn, and that was all.

A gentle wind followed the rain clouds, driving them on northward, a wind that softly clashed the drying corn. A day went by and the wind increased, steady, unbroken by gusts. The dust from the roads fluffed up and spread out and fell on the weeds beside the fields, and fell into the fields a little way. Now the wind grew strong and hard and it worked at the rain crust in the cornfields. Little by little the sky was darkened by the mixing dust, and the wind felt over the earth, loosened the dust, and carried it away. The wind grew stronger. The rain crust broke and the dust lifted up out of the fields and drove gray plumes into the air like sluggish smoke. The corn threshed the wind and made a dry, rushing sound. The finest dust did not settle back to earth now, but disappeared into the darkening sky.

The wind grew stronger, whisked under stones, carried up straws and old leaves, and even little clods, marking its course as it sailed across the fields. The air and the sky darkened and through them the sun shone redly, and there was a raw sting in the air. During a night the wind raced faster over the land, dug cunningly among the rootlets of the corn, and the corn fought the wind with its weakened leaves until the roots were freed by the prying wind and then each stalk settled wearily sideways toward the earth and pointed the direction of the wind.

The dawn came, but no day. In the gray sky a red sun appeared, a dim red circle that gave a little light, like dusk; and as that day advanced, the dusk slipped back toward darkness, and the wind cried and whimpered over the fallen corn.

Men and women huddled in their houses, and they tied handkerchiefs over their noses when they went out, and wore goggles to protect their eyes.

When the night came again it was black night, for the stars could not pierce the dust to get down, and the window lights could not even spread beyond their own yards. Now the dust was evenly mixed with the air, an emulsion of dust and air. Houses were shut tight, and cloth wedged around doors and windows, but the dust came in so thinly that it could not be seen in the air, and it settled like pollen on the chairs and tables, on the dishes. The people brushed it from their shoulders. Little lines of dust lay at the door sills.

In the middle of that night the wind passed on and left the land quiet. The dust-filled air muffled sound more completely than fog does. The people, lying in their beds, heard the wind stop. They awakened when the rushing wind was gone. They lay quietly and listened deep into the stillness. Then the roosters crowed, and their voices were muffled, and the people stirred restlessly in their beds and wanted the morning. They knew it would take a long time for the dust to settle out of the air. In the morning the dust hung like fog, and the sun was as red as ripe new blood. All day the dust sifted down from the sky, and the next day it sifted down. An even blanket covered the earth. It settled on the corn, piled up on the tops of the fence posts, piled up on the wires; it settled on roofs, blanketed the weeds and trees.

The people came out of their houses and smelled the hot stinging air and covered their noses from it. And the children came out of the houses, but they did not run or shout as they would have done after a rain. Men stood by their fences and looked at the ruined corn, drying fast now, only a little green showing through the film of dust. The men were silent and they did not move often. And the women came out of the houses to stand beside their men—to feel whether this time the men would break. The women studied the men's faces secretly, for the corn could go, as long as something else remained. The children stood nearby, drawing figures in the dust with bare toes, and the children sent exploring senses out to see whether men and women would break. The children peeked at the faces of the men and women, and then drew careful lines in the dust with their toes. Horses came to the watering troughs and nuzzled the water to clear the surface dust. After a while the faces of the watching men lost their bemused perplexity and became hard and angry and resistant. Then the women knew that they were safe and that there was no break. Then they asked, What'll we do? And the men replied, I don't know. But it was all right. The women knew it was all right, and the watching children knew it was all right. Women and children knew deep in themselves that no misfortune was too great to bear if their men were whole. The women went into the houses to their work, and the children began to play, but cautiously at first. As the day went forward the sun became less red. It flared down on the dust-blanketed land. The men sat in the doorways of their houses; their hands were busy with sticks and little rocks. The men sat still—thinking—figuring. ∎

Leslie Marmon Silko

FROM "Interior and Exterior Landscapes:
The Pueblo Migration Stories" (1986, 1996)

From a High Arid Plateau in New Mexico

You see that, after a thing is dead, it dries up. It might take weeks or years, but eventually, if you touch the thing, it crumbles under your fingers. It goes back to dust. The soul of the thing has long since departed. With the plants and wild game the soul may have already been born back into bones and blood or thick green stalks and leaves. Nothing is wasted. What cannot be eaten by people or in some way used must then be left where other living creatures may benefit. What domestic animals or wild scavengers can't eat will be fed to the plants. The plants feed on the dust of these few remains.

The ancient Pueblo people buried the dead in vacant rooms or in partially collapsed rooms adjacent to the main living quarters. Sand and clay used to construct the roof make layers many inches deep once the roof has collapsed. The layers of sand and clay make for easy grave digging. The vacant room fills with cast-off objects and debris. When a vacant room has filled deep enough, a shallow but adequate grave can be scooped in a far corner. Archaeologists have remarked over formal burials complete with elaborate funerary objects excavated in trash middens of abandoned rooms. But the rocks and adobe mortar of collapsed walls were valued by the ancient people, because each rock had been carefully selected for size and shape, then chiseled to an even face. Even the pink clay adobe melting with each rainstorm had to be prayed over, then dug and carried some distance. Corncobs and husks, the rinds and stalks and animal bones were not regarded by the ancient people as filth or garbage. The remains were merely resting at a midpoint in their journey back to dust. Human remains are not so different. They should rest with the bones and rinds where they all may benefit living creatures—small rodents and insects—until their return is completed. The remains of things—animals and plants, the clay and stones—were treated with respect, because for the ancient people all these things had spirit and being.

The antelope merely consents to return home with the hunter. All phases of the hunt are conducted with love: the love the hunter and the people have for the Antelope People, and the love of the antelope who agree to give up their meat and blood so that human beings will not starve. Waste of meat or even the thoughtless handling of bones cooked bare will offend the antelope spirits. Next year the hunters will vainly search the dry plains for antelope. Thus, it is necessary to return carefully the bones and hair and the stalks and leaves to the earth, who first created them. The spirits remain close by. They do not leave us.

The dead become dust, and in this becoming they are once more joined with the Mother. The ancient Pueblo people called the earth the Mother Creator of all things in this world. Her sister, the Corn Mother, occasionally merges with her because all succulent green life rises out of the depths of the earth.

Rocks and clay are part of the Mother. They emerge in various forms, but at some time before they were smaller particles of great boulders. At a later time they may again become what they were: dust.

A rock shares this fate with us and with animals and plants as well. A rock has being or spirit, although we may not understand it. The spirit may differ from the spirit we know in animals or plants or in ourselves. In the end we all originate from the depths of the earth. Perhaps this is how all beings share in the spirit of the Creator. We do not know . . .

From the Emergence Place

Pueblo potters, the creators of petroglyphs and oral narratives, never conceived of removing themselves from the earth and sky. So long as the human consciousness remains *within* the hills, canyons, cliffs, and the plants, clouds, and sky, the term *landscape*, as it has entered the English language, is misleading. "A portion of territory the eye can comprehend in a single view" does not correctly describe the relationship between the human being and his or her surroundings. This assumes the viewer is somehow *outside* or *separate from* the territory she or he surveys. Viewers are as much a part of the landscape as the boulders they stand on.

There is no high mesa edge or mountain peak where one can stand and not immediately be part of all that surrounds. Human identity is linked with all the elements of Creation through the clan; you might belong to the Sun Clan or the Lizard Clan or the Corn Clan or the Clay Clan. Standing deep within the natural world, the ancient Pueblo understood the thing as it was—the squash blossom, grasshopper, or rabbit itself could never be created by the human hand. Ancient Pueblos took the modest view that the thing itself (the landscape) could not be improved upon. The ancients did not presume to tamper with what had already been created. Thus *realism,* as we now recognize it in painting and sculpture, did not catch the imaginations of Pueblo people until recently.

[. . .]

The land, the sky, and all that is within them—the landscape—includes human beings. Interrelationships in the Pueblo landscape are complex and fragile. The unpredictability of the weather, the aridity and harshness of much of the terrain in the high plateau country explain in large part the relentless attention the ancient Pueblo people gave to the sky and the earth around them. Survival depended upon harmony and cooperation not only among human beings, but also among all things—the animate and the less animate, since rocks and mountains were known on occasion to move.

The ancient Pueblos believed the Earth and the Sky were sisters (or sister and brother in the post-Christian version). As long as food-family relations are maintained, then the Sky will continue to bless her sister, the Earth, with rain, and the Earth's children will continue to survive. But the old stories recall incidents in which troublesome spirits or beings threaten the earth. In one story, a malicious *ka'tsina,* called the Gambler, seizes the Shiwana, or Rain Clouds, the Sun's beloved children. The Shiwana are snared in magical power late one afternoon on a high mountaintop. The Gambler takes the Rain Clouds to his mountain stronghold, where he locks them in the north room of his house. What was his idea? The Shiwana were beyond value. They brought life to all things on earth. The Gambler wanted a big stake to wager in his games of chance. But such greed, even on the part of only one being, had the effect of threatening the survival of all life on earth. Sun Youth, aided by old Grandmother Spider, outsmarts the Gambler and the rigged game, and the Rain Clouds are set free. The drought ends, and once more life thrives on earth.

Through the Stories We Hear Who We Are

All summer the people watch the west horizon, scanning the sky from south to north for rain clouds. Corn must have moisture at the time the tassels form. Otherwise pollination will be incomplete, and the ears will be stunted and shriveled. An inadequate harvest may bring disaster. Stories told at Hopi, Zuni, and at Acoma and Laguna describe drought and starvation as recently as 1900. Precipitation in west-central New Mexico averages fourteen inches annually. The western pueblos are located at altitudes over 5,600 feet above sea level, where winter temperatures at night fall below freezing. Yet evidence of their presence in the high desert and plateau country goes back ten thousand years. The ancient Pueblo not only survived in this environment, but for many years they also thrived. In A.D. 1100 the people at Chaco Canyon had built cities with apartment buildings of stone five stories high. Their sophistication as sky watchers was surpassed only by Mayan and Incan astronomers. Yet this vast complex of knowledge and belief, amassed for thousands of years, was never recorded in writing.

Instead, the ancient Pueblo people depended upon collective memory through successive generations to maintain and transmit an entire culture, a worldview complete with proven strategies for survival. The oral narrative, or story, became the medium through which the complex of Pueblo knowledge and belief was maintained. Whatever the event or subject, the ancient people perceived the world and themselves within that world as part of an ancient, continuous story composed of innumerable bundles of other stories.

The ancient Pueblo vision of the world was inclusive. The impulse was to leave nothing out. Pueblo oral tradition necessarily embraced all levels of human experience. Otherwise, the collective knowledge and beliefs comprising ancient Pueblo culture would have been incomplete. Thus, stories about the Creation and Emergence of human beings and animals into this world continue to be retold each year for four days and four nights during the winter solstice. The *hummah-hah* stories related events from the time long ago when human beings were still able to communicate with animals and other living things. But beyond these two preceding categories, the Pueblo oral tradition knew no boundaries. Accounts of the appearance of the first Europeans (Spanish) in Pueblo country or of the tragic encounters between Pueblo people and Apache raiders were no more or less important than stories about the biggest mule deer ever taken or adulterous couples surprised in cornfields and chicken coops. Whatever happened, the ancient people instinctively sorted events and details into a loose narrative structure. Everything became a story.

[. . .]

The importance of cliff formations and water holes does not end with hunting stories. As offspring of the Mother Earth, the ancient Pueblo people could not conceive of themselves within a specific landscape, but location, or place, nearly always plays a central role in the Pueblo oral narratives. Indeed, stories are most frequently recalled as people are passing by a specific geographical feature or the exact location where a story took place. The precise date of the incident often is less important than the place or location of the happening. "Long, long ago," "a long time ago," "not too long ago," and "recently" are usually how stories are classified in terms of time. But the places where the stories occur are precisely located, and prominent geographical details recalled, even if the landscape is well known to listeners, often because the turning point in the narrative involved a peculiarity of the special quality of a rock or tree or plant found only at that place. Thus, in the case of many of the Pueblo narratives, it is impossible to determine which came first, the incident or the geographical feature that begs to be brought alive in a story that features some unusual aspect of this location.

There is a giant sandstone boulder about a mile north of Old Laguna, on the road to Paguate. It is ten feet tall and twenty feet in circumference. When I was a child, and we would pass this boulder driving to Paguate village, someone usually made reference to the story about Kochininako,

Yellow Woman, and the Estrucuyo, a monstrous giant who nearly ate her. The Twin Hero Brothers saved Kochininako, who had been out hunting rabbits to take home to feed her mother and sisters. The Hero Brothers had heard her cries just in time. The Estrucuyo had cornered her in a cave too small to fit its monstrous head. Kochininako had already thrown to the Estrucuyo all her rabbits, as well as her moccasins and most of her clothing. Still the creature had not been satisfied. After killing the Estrucuyo with her bows and arrows, the Twin Hero Brothers slit open the Estrucuyo and cut out its heart. They threw the heart as far as they could. The monster's heart landed there, beside the old trail to Paguate village, where the sandstone boulder rests now.

It may be argued that the existence of the boulder precipitated the creation of a story to explain it. But sandstone boulders and sandstone formations of strange shapes abound in the Laguna Pueblo area. Yet, most of them do not have stories. Often the crucial element in a narrative is the terrain—some specific detail of the setting.

A high, dark mesa rises dramatically from a grassy plain, fifteen miles southeast of Laguna, in an area known as Swahnee. On the grassy plain 140 years ago, my great-grandmother's uncle and his brother-in-law were grazing their herd of sheep. Because visibility on the plain extends for over twenty miles, it wasn't until the two sheepherders came near the high, dark mesa that the Apaches were able to stalk them. Using the mesa to obscure their approach, the raiders swept around both ends of the mesa. My great-grandmother's relatives were killed, and the herd was lost. The high, dark mesa played a critical role: the mesa had compromised the safety that the openness of the plains had seemed to assure.

Pueblo and Apache alike relied upon the terrain, the very earth itself, to give them protection and aid. Human activities or needs were maneuvered to fit the existing surroundings and conditions. I imagine the last afternoon of my distant ancestors as warm and sunny for late September. They might have been traveling slowly, bringing the sheep closer to Laguna in preparation for the approach of colder weather. The grass was tall and only beginning to change from green to a yellow that matched the late afternoon sun shining off it. There might have been comfort in the warmth and the sight of the sheep fattening on good pasture that lulled my ancestors into their fatal inattention. They might have had a rifle, whereas the Apaches had only bows and arrows. But there would have been four or five Apache raiders, and the surprise attack would have canceled any advantage the rifles gave them.

Survival in any landscape comes down to making the best use of all available resources. On that particular September afternoon, the raiders made better use of the Swahnee terrain than my poor ancestors did. Thus, the high, dark mesa and the story of the two lost Laguna herders became inextricably linked. The memory of them and their story resides in part with the high, dark mesa. For as long as the mesa stands, people within the family and clan will be reminded of the story of that afternoon long ago. Thus, the continuity and accuracy of the oral narratives are reinforced by the landscape—and the Pueblo interpretation of that landscape is *maintained.*

David Leveson

FROM *A Sense of the Earth* (1972)

At the brink of Canyon de Chelly I met a man searching for a place to live. He wanted a place, he said, connected to the earth, bound to the soil, to rock, where he could earn an honest living and begin to lead a meaningful life. As we chatted, his gaze kept shifting to the bottom of the canyon. There, separated from us by a sheer drop of eight hundred feet, lay brown and green fields and the inconspicuous polygonal mud hogans of Navaho Indians. In shaded recesses at the base of and part way up the cliffs were perched the remains of dwellings built by people who had lived there a thousand years before. The floor and walls of the canyon and the world they contained were graced with a sense of unreality. They seemed forever unreachable—a place, a culture, a history in which we could play no part.

The sight of the canyon and the people living there seemed to invoke a certain remorse in my companion. "For them, it's all right," he said. "They have their place. But for us, there's no place left, or at least I can't find one. Do you know," he asked, "where a man can be at home?"—and left me with that question.

The next morning I was bumping along the floor of the canyon in the back of a four-wheel-drive pickup truck, watching the massive sandstone walls twist past me. Unfettered horses grazed in side canyons; cottonwood groves grew next to where the stream would be when it ran in the wet season. "What place is home?" I asked myself. The lack of a ready answer should have bothered me, but I felt only the shaking of the truck as it rattled on.

In the remoter recesses of the canyon three children were digging for water in the dry stream bed. Like the roots of the cottonwood, they knew where to look.

"What a strange view of the world they must have," I thought. "For them the universe is a narrow strip of sand and mud between massive, sky-

high walls. Within the valley, each rock, each angle and shadow has its special meaning and dimension. Some are places to run to when flash floods
roll down from the east, others are sites reserved for the gods. How at home
they must be here—until rumors of the world above the rim, outside the
canyon disturb them, and life takes them to where the earth is no longer
familiar."

What answer is there to questions such as "where is home?" What
answer can there be? It occurred to me then that the only solution is to be
at home everywhere. But if I had said that to the man looking for a place
to live, he might have thought me unsympathetic; and if I had said the
same thing to the Navaho emerging from the canyon into the modern
world, he might have thought me unfeeling.

Much of life these days, it is true, especially in the cities, is incompat-
ible with the ideal of a real home, and meaningful existence. It is life
divorced from the canyon, from the earth and from other men. But actually the earth is everywhere, and from it, if only we can sense it, there
emanates constantly the wherewithal for man to know what he is and
where he belongs. Awareness of the earth, consciousness of its proximity,
of its inescapable influence—even when not obvious—presents aesthetic
and psychological possibilities largely overlooked or forgotten. Each individual, in canyons or beyond, is deeply affected by his physical surroundings. If it can reach him, knowledge of the earth as reality, rock as material of the universe, landscape as momentary expression of natural process,
is a rich and vital source of sanity and calm for modern man.

It is here that geology and the geologist have a contribution to make.
With the geologist lies the special responsibility and opportunity of
revealing the earth in all its beauty and power. The geologist is in a peculiar position. He has a foot firmly planted in each of two worlds: the
modern scientific world, with all its abstraction and complex tools, and
the world of the actual earth, with its primitive substance and concrete
being. This dual aspect of his profession arms him uniquely, if he so
chooses, to present the earth to modern man in relevant and acceptable
terms. As a scientist, the geologist is part of modern society and thus

speaks its language and is familiar with its problems; at the same time he is a human being in contact with the mystery of the earth and partakes of its nourishment. If geology and the geologist neglect interpretation of the earth to society, they are guilty of relinquishing what should be one of their major contributions. ▪

N. Scott Momaday

FROM "An American Land Ethic" in *The Man Made of Words* (1997)

II

Once in his life a man ought to concentrate his mind upon the remembered earth, I believe. He ought to give himself up to a particular landscape in his experience, to look at it from as many angles as he can, to wonder about it, to dwell upon it. He ought to imagine that he touches it with his hands at every season and listens to the sounds that are made upon it. He ought to imagine the creatures there and all the faintest motions of the wind. He ought to recollect the glare of noon and all the colors of the dawn and dusk.

The Wichita Mountains rise out of the southern plains in a long crooked line that runs from east to west. The mountains are made of red earth, and of rock that is neither red nor blue but some very rare admixture of the two, like the feathers of certain birds. They are not so high and mighty as the mountains of the Far West, and they bear a different relationship to the land around them. One does not imagine that they are distinctive in themselves, or indeed that they exist apart from the plain in any sense. If you try to think of them in the abstract, they lose the look of mountains. They are preeminently an expression of the larger landscape, more perfectly organic than one can easily imagine. To behold these mountains from the plain is one thing; to see the plain from the mountains is something else. I have stood on the top of Mount Scott and seen the earth below, bending out into the whole circle of the sky. The wind runs always close upon the slopes, and there are times when you hear the rush of it like water in the ravines.

Here is the hub of an old commerce. More than a hundred years ago the Kiowas and Comanches journeyed outward from the Wichitas in every direction, seeking after mischief and medicine, horses and hostages. Sometimes they went away for years, but they always returned, for the land had got hold of them. It is a consecrated place, and even now there is something of the wilderness about it. There is a game preserve in the hills. Animals graze away in the open meadows or, closer by, keep to the shadows of the groves: antelope and deer, longhorns and buffalo. It was here, the Kiowas say, that the first buffalo came into the world.

The yellow grassy knoll that is called Rainy Mountain lies a short distance to the north and west. There, on the west side, is the ruin of an old school where my grandmother went as a wild girl in blanket and braids to learn of numbers and of names in English. And there she is buried.

Most is your name the name of this dark stone.
Deranged in death, the mind to be inheres
Forever in the nominal unknown,
The wake of nothing audible he hears
Who listens here and now to hear your name.

The early sun, red as a hunter's moon,
Runs in the plain. The mountain burns and shines;
And silence is the long approach of noon
Upon the shadow that your name defines—
And death this cold, black density of stone.

III

I am interested in the way that a man looks at a given landscape and takes possession of it in his blood and brain. For this happens, I am certain, in the ordinary motion of life. None of us lives apart from the land entirely; such an isolation is unimaginable. We have sooner or later to come to terms with the world around us—and I mean especially the physical world, not only as it is revealed to us immediately through our senses, but also as it is perceived more truly in the long turn of seasons and of years. And we must come to moral terms. There is no alternative, I believe, if we are to realize and maintain our humanity, for our humanity must consist in part in the ethical as well as in the practical ideal of preservation. And particularly here and now is that true. We Americans need now more than ever before—and indeed more than we know—to imagine who and what we are with respect to the earth and sky. I am talking about an act of the imagination, essentially, and the concept of an American land ethic.

It is no doubt more difficult to imagine the landscape of America now, than it was in, say, 1900. Our whole experience as a nation in this century has been a repudiation of the pastoral ideal which informs so much of the art and literature of the nineteenth century. One effect of the technological revolution has been to uproot us from the soil. We have become disoriented, I believe; we have suffered a kind of psychic dislocation of ourselves in time and space. We may be perfectly sure of where we are in relation to the supermarket and the next coffee break, but I doubt that any of us knows where he is in relation to the stars and to the solstices. Our sense of the natural order has become dull and unreliable. Like the wilderness itself, our sphere of instinct has diminished in proportion as we have failed to imagine truly what it is. And yet I believe that it is possible to formulate an ethical idea of the land—a notion of what it is and must be in our daily lives—and I believe moreover that it is absolutely necessary to do so.

It would seem on the surface of things that a land ethic is something that is alien to, or at least dormant in, most Americans. Most of us have developed an attitude of indifference toward the land. In terms of my own experience, it is difficult to see how such an attitude could ever have come about. ■

Wide enough to keep you looking

Open enough to keep you moving

Dry enough to keep you honest

Prickly enough to make you tough

Green enough to go on living

Old enough to give you dreams

—Gary Snyder, "Earth Verse"
in *Mountains and Rivers
Without End* (1996)

Contributors

Mary Austin (1868–1934) wrote fiction and nonfiction. Born in Illinois, she moved to California when she was twenty. She settled in Inyo County, where she and her then husband were involved in the California water wars. Her best-known book, *The Land of Little Rain* (1903), is an account of life in the California desert.

Kofi Awoonor (b. 1935) is a Ghanaian novelist and poet. His grandmother was a dirge singer, and his early work combines this oral poetry and religious symbolism to depict Africa during decolonization. Awoonor earned a Ph.D. in literature from the State University of New York, Stony Brook, in 1972, while in political exile from his native country. He returned to Ghana in 1975, and in recent years he has focused on political activities.

Wendell Berry (b. 1934) is the author of more than thirty books of poetry, essays, and fiction. He lives and farms on the family land where he was born at Port Royal, Kentucky. He is an eloquent defender of family, rural communities, and traditional family farms. Among his titles are *The Unsettling of America*, *The Distant Land*, and *Another Turn of the Crank*.

Isabella Bird (1831–1904) was an English traveler and writer. For over four decades, her adventures in such places as North America, China, India, Tibet, Turkey, Persia, and Kurdistan were featured in journals and magazines. She also chronicled her travels in a number of books, including *A Lady's Life in the Rocky Mountains* (1879).

Thomas Burnet (c. 1635–1715) was royal chaplain to King William III of England and a scientist. He published *The Sacred Theory of the Earth* (1681–1689) in which he tried to reconcile what he observed of geology during his travels in Europe with his understanding of the Bible.

John Calderazzo is a professor at Colorado State University in Fort Collins, where he teaches nonfiction writing and literature. His articles on a range of topics, including natural history, the relationship between science and culture, Buddhism, and Asia, have appeared in the *Georgia Review, Audubon, Orion,* and other publications. *Rising Fire: Volcanoes and Our Inner Lives* was published in 2004.

Rachel Carson (1907–1964), a writer and ecologist, worked for the U.S. Fish and Wildlife Service. In 1952 she published her best-selling study of the ocean, *The Sea Around Us,* which was followed by *The Edge of the Sea* (1955). In *Silent Spring* (1962) she challenged the practices of agricultural scientists and the government and called for a change in the way humankind viewed the natural world. Her work was a catalyst for the modern environmental movement.

Kamo no Chomei (1153–1216) was born in Kyoto, Japan, into a family of Shinto priests. After a career as a poet in the imperial court, he gave up Shintoism and became a Buddhist monk, spending much of his later life as a hermit. His essay "An Account of My Hut" (*Hojoki*) extols the virtues of living a simple, rural life.

Lucille Clifton (b. 1936) is an actress and poet. Her poems are celebrations of African American heritage, culture, and history. She has also written award-winning books for children, and she served as Poet Laureate of Maryland from 1979 to 1982.

Hans Cloos (1885–1951) was a German geologist. A pioneer in the study of granite tectonics (the deformation of crystalline rocks), he also studied the structure and development of the continents. He was a professor at the Universities of Breslau and Bonn.

Charles Darwin (1809–1882) was a British naturalist, geologist, and writer. Based on the data he collected during his five-year voyage on the *Beagle,* he developed his own understanding of diversification in nature, the transmutation of species, and natural selection. He published his ideas in 1859 in *The Origin of Species.*

Jan DeBlieu has written for the *New York Times Magazine, Audubon,* and *Orion.* She won the 1999 John Burroughs Medal for Distinguished Natural History Writing for *Wind: How the Flow of Air Has Shaped Life, Myth, and the Land.* She lives on the Outer Banks of North Carolina.

Annie Dillard (b. 1945) won the Pulitzer Prize for general nonfiction in 1975 with her first book of prose, *Pilgrim at Tinker Creek.* Her other work includes *Teaching a Stone to Talk* (1982), *An American Childhood* (1987), and *For the Time Being* (1999).

Ivan Doig (b. 1939), born into a family of Montana ranch hands and sheep-herders, worked as a journalist before turning full time to literary writing, including his Montana trilogy, *English Creek*, *Dancing at the Rascal Fair*, and *Ride with Me, Mariah Montana*.

Rita Dove (b. 1952) is Commonwealth Professor of English at the University of Virginia in Charlottesville. She served as Poet Laureate of the United States from 1993 to 1995 and as Poet Laureate of the Commonwealth of Virginia. The recipient of many literary and academic honors, including the 1987 Pulitzer Prize in Poetry and the 1996 National Humanities Medal, she has published poetry, short stories, essays, and a novel.

Loren Eiseley (1907–1977) was an anthropologist, environmentalist, poet, and writer. For many years he taught at the University of Pennsylvania, where he was head of the anthropology department. Many of his essays examine the history of civilization and our relationship with the natural world. His best-known book is *The Immense Journey* (1957).

E. M. Forster (1879–1970) was an English novelist. His travels to Egypt, Germany, and India influenced his writing. In *Howard's End* (1910) and *A Passage to India* (1924) he explores the irreconcilability of class differences.

August Gansser (b. 1910) played an eminent role in determining the geological structures of the Himalaya and explaining the origin of the Asiatic mountain belts. *Geology of the Himalaya* (1964) has become a classic in descriptive geology. In 1936 he took part in the first Swiss expedition to the Himalaya, traveling by foot and mule. He is professor emeritus of geology at the Federal Institute of Technology in Zurich, Switzerland.

Sir Archibald Geikie (1835–1924) was a prominent Scottish geologist and author. He was the first director of the Geological Survey established for Scotland and concurrently served as professor of geology and mineralogy at the University of Edinburgh. He headed the Geological Survey of Great Britain from 1881 to 1901.

Grove Karl Gilbert (1843–1918) was the first geologist to realize that the craters on the moon were formed by meteors. His work in geology was seminal in a number of areas, from sediments to the effects of hydraulic mining. In 1879 he was appointed senior geologist in the newly established United States Geological Survey, where he served until his death.

Ray Gonzalez is professor of English at the University of Minnesota. His work includes ten volumes of poetry, two collections of short stories, and a collection of essays, *Underground Heart* (2002).

Stephen Jay Gould (1941–2002) was an American paleontologist, evolutionary biologist, and historian of science. He was a professor of geology and zoology at Harvard University from 1967 until his death. He wrote many popular, award-winning books on science, including *Wonderful Life* (1991), about the Burgess shale fossils, and the essay collections *The Panda's Thumb* (1980) and *The Flamingo's Smile* (1985).

Han-shan, or "Cold Mountain" (c. 627–c. 650), was a Chinese hermit who lived during the T'ang dynasty. Some three hundred of his poems have survived the centuries, of which the poet Gary Snyder has translated several into English.

Jacquetta Hawkes (1910–1996) was a British archeologist. As archeology correspondent for the London *Times* and through such books as *The First Great Civilizations* (1973), *The Atlas of Ancient Archaeology* (1975), and *A Land* (1951), she popularized the subject for the general public.

Arnold Heim (1882–1965) was a Swiss petroleum geologist. In 1936 he and fellow geologist August Gansser made one of the first geologic expeditions to the northwestern Himalaya.

Daniel Hillel is professor emeritus of soil, plant, and environmental sciences at the University of Massachusetts, Amherst. An environmental scientist and hydrologist, he has worked throughout the Middle East as a consultant to governments and international organizations. His publications include scientific reports, popular articles, and nineteen books.

Jane Hirshfield (b. 1953) is an essayist, translator, and poet. Her works include *Given Sugar, Given Salt* (2001), *Nine Gates: Entering the Mind of Poetry* (1997), and *The Ink Dark Moon* (1990), translations of two women of the ancient court of Japan.

Garrett Hongo (b. 1951) is a poet and essayist, and professor of creative writing at the University of Oregon, Eugene. Of Japanese heritage, he was born in Hawaii and raised in Los Angeles. In both his poetry and a memoir, *Volcano* (1995), he explores the landscape of his birthplace and the Japanese American experience. *The River of Heaven* (1988) was a finalist for the Pulitzer Prize.

James D. Houston (b. 1933) is an author and essayist. He has written seven novels, including the award-winning *The Last Paradise* (1998) and *Snow Mountain Passage* (2001). He coauthored, with his wife, Jeanne Wakatsuki Houston, *Farewell to Manzanar* (1973), which tells the story of her family's internment during World War II.

Langston Hughes (1902–1967) was among the best known poets of the Harlem Renaissance, the 1920s and 1930s flowering in African-American literature, arts, and culture. Hughes was also a playwright, novelist, essayist, editor, journalist, lyricist, translator, and author of short stories and children's books. Among his many works are the novel *Not Without Laughter* (1930), poetry volumes *The Weary Blues* (1926) and *Montage of a Dream Deferred* (1951), and the play *Mulatto* (1935).

James Hutton (1726–1797), a Scottish geologist, chemist, and naturalist, is considered the father of modern geology. He is noted for formulating uniformitarianism, the idea that existing processes acting in the same way as present on Earth's surface are sufficient to explain all past geological changes. He also postulated that Earth must be much older than the Bible allows. His ideas ultimately influenced Charles Darwin's theories of evolution.

John Imbrie (b. 1925), professor emeritus at Brown University, is one of the founders of modern paleo-oceanography. He pioneered the use of computers to analyze microscopic marine fossil data; his work on Pleistocene marine sediments eventually paved the way for explaining Earth's great ice ages. His book *Ice Ages: Solving the Mystery* (1979), coauthored with his daughter, Katherine Palmer Imbrie, popularizes the story of this quest.

Katherine Palmer Imbrie is a features writer for the *Providence Journal.* With her father, John Imbrie, she wrote *Ice Ages: Solving the Mystery* (1979).

James Joyce (1882–1941) was an Irish novelist. He is noted for his experimental use of language and stream-of-consciousness technique in such works as *Ulysses* (1922) and *Finnegans Wake* (1939).

Nikos Kazantzakis (1883–1957) was a Greek writer and philosopher. His novels include *The Odyssey: A Modern Sequel* (1938), *Zorba the Greek* (1946), and *The Last Temptation of Christ* (1955). He also published a memoir, *Report to Greco* (1961).

William Langewiesche (b. 1955) is a pilot, travel writer, and journalist. His books include *American Ground: Unbuilding the World Trade Center* (2002) and *The Outlaw Sea* (2004). He is a regular contributor to the *Atlantic Monthly* and has twice won the National Magazine Award for Excellence in Reporting.

Ursula K. Le Guin (b. 1929) is an author whose work includes poetry, children's books, novels, short stories, essays, and translations. She is best known for her works of science fiction and fantasy, such as *Left Hand of Darkness* (1969) and *The Dispossessed* (1974).

David Leveson (b. 1934) is professor of geology at Brooklyn College, City University of New York. His current research is on key figures in the history of geology and the development of online courses in geology.

Barry Lopez (b. 1945) is the author of fourteen books of fiction and nonfiction. His works include the National Book Award–winning *Arctic Dreams* (1986), *Of*

Wolves and Men, for which he received the John Burroughs Medal, the essay collection *About This Life* (1999), and *Resistance* (2005), a work of fiction. He has received the Award in Literature from the American Academy of Arts and Letters, a Lannan Award, and many other honors.

Audre Lorde (1934–1992) was an American writer and activist who published a dozen volumes of poetry and ten works of prose. She served as Poet Laureate of New York from 1991 to 1993. She called herself "black, lesbian, mother, warrior, poet."

Henno Martin (1910–1998) was a German geologist. In 1935 he and a coworker went to South-West Africa (now Namibia) to do fieldwork; when World War II broke out, rather than being held in an internment camp, they fled into the Namib desert, where they remained for two and a half years. Martin wrote about this experience in *The Sheltering Desert* (1957).

John McPhee (b. 1931) is a writer of nonfiction. A frequent contributor to the *New Yorker*, he has published twenty-nine books on subjects as diverse as oranges, the basketball player Bill Bradley, the Alaskan wilderness, and the shad. *Annals of the Former World* (1998), which brings four previously published books on the geology of North America under one cover, won the Pulitzer Prize in 1999.

N. Scott Momaday (b. 1934) is a Native American author. His first novel, *House Made of Dawn* (1968), won the Pulitzer Prize in 1969. His most recent book, *In the Bear's House* (1999), a collection of prose, poetry, and paintings, is a testament to his Kiowa heritage. Momaday retired as professor of English at the University of Arizona, Tucson.

John Muir (1838–1914) was a traveler, writer, inventor, naturalist, and preservationist. Born in Scotland, he came to the United States with his family in 1849 and eventually made his way to California. Captivated by the Sierra Nevada, he began to study the geology of the range. When he realized that Yosemite was threatened by grazing, he worked to have it declared a national park. He helped found the Sierra Club in 1892 and was elected its first president, a position he held until his death. He wrote over three hundred articles and ten major books recounting his travels and expounding his naturalist philosophy.

Hikaru Okuizumi (b. 1956) is a Japanese novelist. His first novel to be published in English, *The Stones Cry Out* (1993), won the Akutagawa Prize.

Michael Ondaatje (b. 1943), born in Sri Lanka, is a Canadian author and editor. In addition to memoir and poetry, he has written several works of fiction, including *The English Patient* (1992), winner of the Booker Prize, and *Anil's Ghost* (2001).

Alfonso Ortiz (1939–1997) was born in the San Juan Pueblo, New Mexico. He is best known for his book *The Tewa World: Space, Time, Being, and Becoming in a Pueblo Society* (1969). Among many honors and awards, in 1982 he received a MacArthur Fellowship. He was a professor of anthropology at the University of New Mexico.

Louis Owens (1948–2002) was a critical interpreter of Native American literature, a Steinbeck scholar, and a fiction and nonfiction writer. In much of his writing, Owens, who considered himself a mixed-blood American (Choctaw, Cherokee, and Irish-American), explored the dilemmas of being from multiple heritages.

Pliny the Younger (c. A.D. 61–c. 112), properly Gaius Plinius Caecilius Secundus, was a Roman senator under the emperors Domitian, Nerva, and Trajan, and is best known for his ten books of letters, which are an important source for Roman social history. He was brought up by his uncle, Pliny the Elder, commander of the fleet at Misenum, on the Bay of Naples.

John Wesley Powell (1834–1902) was a U.S. soldier, geologist, and explorer of the American West. He was professor of geology at Illinois Wesleyan University and helped found the Illinois Museum of Natural History. His 1869 surveying trip down the Green and Colorado rivers included the first passage through the Grand Canyon. He later became director of the U.S. Geological Survey.

Stephen J. Pyne is a professor in the School of Life Sciences at Arizona State University, where he specializes in the history of ecology and the history of exploration. He is one of the world's foremost experts on the natural history of fire, with a dozen books on the subject, most recently *Tending Fire* (2004). He received a MacArthur Fellowship in 1988.

John Ruskin (1819–1900) was a British poet, artist, art critic, and social theorist. A preeminent Victorian intellectual, for ten years he was Slade Professor of Art at Oxford University. His works include *The Stones of Venice* (1851–1853), *The Seven Lamps of Architecture* (1859), the five-volume *Modern Painters* (1843–1860), and *Unto This Last* (1862).

Antoine de Saint-Exupéry (1900–1944) was a French writer and aviator. He pioneered international postal flights between Europe and Africa and later in South America. During World War II, while flying what was to have been his last reconnaissance mission for the Allies, his plane disappeared. The wreckage was not found until 2004. Saint-Exupéry's books include *Vol de nuit* (Night Flight, 1931) and *Le Petit Prince* (The Little Prince, 1943).

Nanao Sakaki (b. 1923) is known as the "godfather of Japanese hippies." After World War II he became a wandering scholar and poet, and in the 1960s he befriended Allen Ginsberg and Gary Snyder, who invited him to visit the United States. In addition to writing his own poetry, he has translated the poems of Issa Kobayashi, one of the masters of the haiku form.

Eliza Ruhamah Scidmore (1856–1928) was a travel writer and photographer. She worked for many years for the National Geographic Society and spent long periods in Asia. She published several books, including *Java, the Garden of the East* (1897), and *As The Hague Ordains* (1907).

William Scoresby, Jr. (1789–1857) was an English sea captain, Arctic explorer, scientist, and clergyman. While the captain of a whaling vessel, he charted the coast of Greenland and for seventeen years studied the Arctic Ocean, aiding his government's decision to search for new polar routes.

Vikram Seth (b. 1952), a native of India, is best known for his novels, which include *A Suitable Boy* (1993) and *An Equal Music* (1999). He has also written poetry, librettos, memoir, a book for children, and the travel book *From Heaven Lake: Travels Through Sinkiang and Tibet* (1983).

Paul Shepard (1925–1996) was a human ecologist and an influential thinker whose work helped shape the environmental movement. His books include *The Tender Carnivore and the Sacred Game* (1973), *Thinking Animals: Animals and the Development of Human Intelligence* (1978), *Nature and Madness* (1982), and *The Others: How Animals Made Us Human* (1997).

Leslie Marmon Silko (b. 1948) is the author of novels, short stories, essays, poetry, articles, and screenplays. Of Anglo, Mexican, and Native American heritage, she was raised on the Laguna Pueblo of New Mexico and incorporates the stories and sensibilities of her people in her work. Her most recent book is *Gardens in the Dunes* (1999). She received a MacArthur Fellowship in 1981.

Gary Snyder (b. 1930) is a poet, essayist, lecturer, and environmental activist. His honors include a Pulitzer Prize for *Turtle Island* (1974) and the 1997 Bollingen Prize for Poetry. In 1996 he published *Mountains and Rivers Without End*, a work written over a forty-year period, and in 2004 he published *Danger on Peaks*.

Luther Standing Bear (1868?–1939) was an Oglala Lakota. During his life, he hunted buffalo as a child, attended a boarding school, joined Buffalo Bill's Wild West Show as an interpreter and performer, started a dry-goods store, became chief of his tribe, acted in films, and wrote four books, including *My People, the Sioux* (1928) and *Land of the Spotted Eagle* (1933).

John Steinbeck (1902–1968) was awarded the Nobel Prize for Literature in 1962. His books, social novels dealing with the economic problems of rural labor, include *The Grapes of Wrath* (1939), which won a Pulitzer Prize, and *East of Eden* (1952).

Luci Tapahonso (b. 1953), a Navajo originally from Shiprock, New Mexico, is a poet, storyteller, scholar, and Native advocate. She is professor of English at the University of Arizona in Tucson, where she teaches poetry writing and American Indian literature. She is the author of five books of poetry and three children's books.

Haroun Tazieff (1914–1998), born in Warsaw, was a prominent French vulcanologist. In 1958 he became director of the Laboratory of Vulcanology of the Institut de Physique du Globe, Paris. From 1988 to 1995 he served as president of a French national committee assessing major volcanic risks. He wrote several books and made numerous documentary films about volcanoes.

Mark Twain (1835–1910), born Samuel Clemens, was a riverboat pilot, newspaperman, author, and lecturer. He wrote dozens of works of fiction and nonfiction, including *Life on the Mississippi* (1883), *The Adventures of Huckleberry Finn* (1884), and *Letters from the Earth* (1909).

Ofelia Zepeda (b. 1954) is a professor in the departments of Linguistics and American Indian Studies at the University of Arizona, Tucson. A member of the Tohono O'odham Nation of southern Arizona, she published the first teaching grammar of O'odham, *A Papago Grammar* (1983). In 1999 she received a MacArthur Fellowship for her work in American Indian language education and preservation.

Ann Zwinger (b. 1925) is an author and naturalist. She has long taught writing and Southwest studies at Colorado College in Colorado Springs. Among her eighteen natural history books are *Land Above the Trees* (1989), *Down Canyon* (1995), *The Nearsighted Naturalist* (1998), and *Run, River, Run* (1984), winner of the John Burroughs Medal for Nature Writing.

Permissions

Photo Credits

Part 1, page xviii: River-sculpted Vishnu Schist, Grand Canyon National Park, Arizona. Photo by Lauret Savoy.

Part 2, page 42: Double Arch, Arches National Park, Utah. Photo by Lauret Savoy.

Part 3, page 70: Aerial view of the San Andreas Fault and offset streams on the Carrizo Plain, California. Photo by Michael Collier.

Part 4, page 98: Mount St. Helens eruption of May 18, 1980. Photo by Austin Post.

Part 5, page 154: Rock glacier, Grand Tetons, Wyoming. Photo by John Shelton.

Part 6, page 192: Tukumm Creek cuts Marble Canyon, Kootenay National Park, southeastern British Columbia. Photo by Lauret Savoy.

Part 7, page 228: Multiple medial moraines, Barnard glacier, Alaska. Photo by Austin Post.

Part 8, page 254: Star Dunes, Death Valley, California. Photo by John Shelton.

Part 9, page 288: Ancestral Pueblo ruins, Canyon de Chelly, Arizona. Photo by Lauret Savoy.

Photographs by Austin Post and John Shelton were provided by EPIC—the Easterbrook Photo/Image Center.